Praise for *The Human Fc*

"This is an ambitious and timely book that builds
historical research. Key attractions include the cor
tion of teaching in many world history courses and the embrace not only of big
changes like the advent of agriculture or industrialization, but less familiar
developments such as the environmental impact of migration patterns. The
result is a real sense of how humans have interacted with nature and the ways
current environmental issues connect to the past."
Peter N. Stearns, George Mason University,
author of *The Industrial Revolution in World History*

"Anthony Penna is to be congratulated for producing one of the first environmental histories to embrace the entire world and all of human history. As our
relationship with the world's diverse environments deteriorates such educa-
tional resources are becoming increasingly vital."
David Christian, Macquarie University, formerly of SDSU,
author of *Maps of Time: An Introduction to Big History*

"Anthony Penna's *The Human Footprint* is an insightful survey of global history,
one distinguished by its perceptive interweaving of human and natural history.
In clear and accessible prose, it provides a masterful synthesis of scholarship
across a wide range of disciplines. Its breadth and sophistication—and its rele-
vance to the world today—make it a compelling read."
Jeffrey K. Stine, Curator of Engineering and Environmental History,
Smithsonian Institution, author of *America's Forested Wetlands:
From Wasteland to Valued Resource*

"Anthony Penna weaves human and natural history together into a single,
compelling story. In his vision, human innovation, culture and exchange, nutrition, atmospheric chemistry and plate tectonics are just a few of the many processes that come together in an endless dance of engagement and change. Penna
synthesizes recent research that shows how human biology and culture are the
products of natural history, and also uses the methods of human history to
show how the natural environment has developed through unique and specific
events that now include the impact of humans themselves. The pasts and the
fates of humanity, nature and the Earth are one and the same."
Adam McKeown, Columbia University, author of *Melancholy Order: Asian
Migration and the Globalization of Borders 1834–1929*

"Combining wide knowledge with an eye for the essential, Penna takes a truly
vast and challenging subject—the natural and human history of the earth—and
distills it into a volume that is reliable, accessible, and illuminating."
William B. Meyer, Colgate University,
author of *Human Impact on the Earth*

The Human Footprint
A Global Environmental History

Anthony N. Penna

A John Wiley & Sons, Ltd., Publication

This edition first published 2010
© 2010 Anthony N. Penna

Blackwell Publishing was acquired by John Wiley & Sons in February 2007. Blackwell's publishing program has been merged with Wiley's global Scientific, Technical, and Medical business to form Wiley-Blackwell.

Registered Office
John Wiley & Sons Ltd, The Atrium, Southern Gate, Chichester, West Sussex, PO19 8SQ, United Kingdom

Editorial Offices
350 Main Street, Malden, MA 02148-5020, USA
9600 Garsington Road, Oxford, OX4 2DQ, UK
The Atrium, Southern Gate, Chichester, West Sussex, PO19 8SQ, UK

For details of our global editorial offices, for customer services, and for information about how to apply for permission to reuse the copyright material in this book please see our website at www.wiley.com/wiley-blackwell.

The right of Anthony N. Penna to be identified as the author of this work has been asserted in accordance with the Copyright, Designs and Patents Act 1988.

Library of Congress Cataloging-in-Publication Data

Penna, Anthony N.
  The human footprint : a global environmental history / Anthony N. Penna.
    p.   cm.
  Includes bibliographical references and index.
  ISBN 978-1-4051-8772-5 (hardcover : alk. paper) – ISBN 978-1-4051-8771-8 (pbk. : alk. paper)
1. Human ecology–History. 2. Ecology–History. 3. Climatic changes–History. 4. Natural history. I. Title.
  GF13.P46 2010
  304.2–dc22

                                                                     2009008319

A catalogue record for this book is available from the British Library.

Set in 10/13pt Palatino
by SPi Publisher Services, Pondicherry, India
Printed in Singapore by Fabulous Printers Pte Ltd

01   2010

In Memoriam

To my grandparents whose life journeys created opportunities

Vincenzo Penna (1870–1950)
Filomena Penna (1873–1956)
Niccolo DePalma (1888–1916)
Carolina DePalma Bosco (1890–1981)
Pellegrino Bosco (1882–1963)

To my parents who provided incentives

Anthony Penna (1907–1987)
Mary DePalma Penna (1914–2003)

CONTENTS

Contents

Contents

Contents

LIST OF FIGURES

List of Figures

ACKNOWLEDGMENTS

Many of the global challenges facing humans in the twenty-first century will require knowledge about the complexity of the planet's climate system and contextually informed understandings about our role as stewards of its fragile ecology. For educators at all levels of teaching, the task of integrating ecology and environmental history into local, regional, and national histories will become a compelling endeavor requiring flexibility, adaptation, and new knowledge. Survey courses in global environmental history and specialized national and regional environmental histories will become important dimensions of the evolving history and social science curriculum and the growing field of environmental studies.

This book is the result of two journeys; the first took place in numerous conversations with my wife-to-be, Channing, as we walked the nearly 500-mile pilgrimage route across northern Spain, *el Camino de Santiago*, in September 2001. The second journey, familiar to all writers, begins with the assertion that this subject is worth writing about.

From 2002 until publication in 2010, the research took me on its own journey and became a central part of each day. In writing this book, I relied on the knowledge, insight, and wisdom of many scholars in various disciplines. I also had the benefit of many colleagues who read earlier versions of single chapters, a cluster of chapters, and the entire manuscript during its initial stages of development. My former teacher, colleague at Carnegie-Mellon University, and long-term friend, Irving H. Bartlett, the John F. Kennedy Professor Emeritus of American Civilization at the University of Massachusetts, Boston, remained an enthusiastic critic and supporter of this work. Irving's death in July 2006 denied us the pleasure of celebrating its publication over lunch after a marathon bike ride along the commonwealth's southeastern coast. Joel A. Tarr, the Richard S. Caliguiri University Professor at Carnegie-Mellon University, also a former teacher, colleague and long-term friend, gave unselfishly of his time and knowledge as this project unfolded.

I want to thank Martin Melosi, Jeffrey Stine, George Dehner, and Jason Eden who read earlier versions of some chapters, and Jeffrey and George along with Professors Qingjia Edward Wang, Rowan and Beijing Universities, and Eric Vanhaute of the University of Ghent in Belgium and the many anonymous reviewers from the United States, Canada, and the United Kingdom who read the entire manuscript. Their detailed critiques helped me to refocus, revise, and rewrite.

From the outset, I received the invaluable aid of my research assistants. Michael Mezzano and David Adams compiled the initial bibliographies while Jeanine Rees, Katherine Platt, and Eric Skidmore located materials for chapters on manufacturing, industrialization, consumption, and energy. During the final year of the project, Neysa King and Colin Sargent conducted electronic searches for photographs, charts, graphs, and maps. The fruit of their labor is found throughout this book. In addition, Colin used his advanced technical skills to turn all graphic material into electronic files and format the entire manuscript in accordance with the publisher's guidelines. Elizabeth Somerset's electronic index made the final preparation of a published index less onerous. All three contributed mightily at the end and I am in their debt. I want to thank Peter Coveney, executive editor, Deidre Ilkson, developmental editor, and Galen D. Smith, editorial assistant, for their patience, advice, and judicious editing.

Undergraduate and graduate students enrolled in my global environmental history courses and seminars and came from majors in history, political science, sociology, economics, international relations, biology, civil and environmental engineering, chemistry, and environmental studies. They brought with them knowledge and ideas that informed our ongoing conversations. Many entered these courses skeptical about the importance of historical perspective in understanding present environmental conditions. All left with a willingness to read more widely in other disciplines and with a richer, contextualized understanding of the environmental costs of human action.

Also, I have gained much insight from two seminars on world history. One is an ongoing joint series that began in 2007–8 between

Tufts University and Northeastern University, directed by Felipe Fernandez-Armesto at Tufts and Laura Frader at Northeastern. Sven Beckert, Harvard University, and Dominic Sachsenmaier, Duke University, directed the other, a weekend conference titled "Global History, Globally," in February, 2008. Both the series and the conference provided papers, presentations, and discussions that penetrated more parochial boundaries and encouraged participants to engage in broader historical and comparative perspectives.

While deeply involved in her own creative enterprise, my wife, Channing, remained unflinching in her enthusiasm and support for my work.

INTRODUCTION

The Nature of World History

There has been a surging interest in the field of world history as scholars attempt to cross regional and national boundaries. By weaving national biographies into world history, we engage the fragments of history by fitting them together in a coherent and meaningful whole. Herein lies one of the big intellectual challenges for world historians, namely "to transcend national frontiers, and study forces such as population movements, economic fluctuations, climatic changes, and the transfer of technologies."[1] It remains a vibrant yet contested subject not only because it disturbs the 200-year dominance of European and American national narratives and a trend of subordinating the rest of the world to a Euro-American paradigm but also because no institutional framework in the profession of historians exists for world history. As a highly successful approach for studying change over time, the nation state will remain the primary vehicle for analyzing the experiences of individual, local, and regional communities. For world historians, "exploring the connections, comparisons, and systems that help to situate historical development in larger appropriate contexts"[2] will increasingly become their domain.

Micro-historical studies that focus on vertical trends dominate the research agendas of most American and European universities. The continuing importance of specific studies will remain unchanged but recent developments in world history argue for what one historian described as a horizontally integrative macro-history, one that described and explained interrelated historical phenomena.[3] Looking inward to describe changes in aspiring national histories and outward in an effort to make inter-regional connections and exchanges places less emphasis on persons and nations. As historians go about this transformation in describing world historical events, they attempt to de-center Europe, despite its dominance in world affairs from the beginning of industrialization through World War II.

By de-centering Europe, other parts of the world come into clearer focus in the millennia before industrialization. The coal revolution released energy from the fossil remains of plants and animals, altered global material culture in the nineteenth century, and served as a point of demarcation, separating the pre-modern from the modern world. Additionally, world history encompasses a longer and deeper history, one that is not bounded within a Judeo-Christian framework and one in which the fossilized remains of humans allow archeologists, paleo-anthropologists, and geneticists to gain insight into the physiological development of early humans. Their diets, health, and vulnerability to disease and predation serve as windows into a deep history of humanity.[4]

The Nature of World Environmental History

This world environmental history is not a comprehensive survey of human history focused on civilizations, nations, personal biographies, or events. Plenty of existing world histories accomplish this important task. This one focuses on great transformations in human history and the relationship between human history and natural history, with the emphasis on human world/natural world interrelationships. In the words of J. R. McNeill, it "concerns itself with changes in biological and physical environments, and how those changes affect human societies."[5] It approaches this challenge by reformulating prior knowledge about local, regional, and national histories and by using knowledge that integrates and interrelates these histories.

Additionally, some of the reformulation occurs by crossing disciplinary boundaries and using the research findings and new knowledge of geologists, climatologists, evolutionary biologists, archeologists, paleo-anthropologists, demographers, economists, and the social sciences as they study earth history, evolution, agricultural productivity, urban planning, manufacturing, industry, consumption, and energy use. Scholars across the disciplines are committed to understanding developments as they have been and are now in the real world, not as imagined constructions without a scientific basis. Each discipline, with its own perspective, contains a unique theoretical structure and proof process. In essence, scholars may study the same subject from their traditional disciplinary perspectives but arrive at different places studying the same "real world." In its own way, this book integrates great global transformations and evolutions, telescoped from planetary and human origins to modern consumption patterns and energy uses. Each effort to integrate disciplines as well as local, regional, national, and transnational histories makes the global transformations that draw on ecology and environment possible.

Earth History and Human Origins

Because histories of humankind assume a global ecological system that made life possible and sustainable, they ignore the origins of the Earth and its development over billions of years. Leaving little about Earth history to the imagination, this book begins with an examination of the evolutionary history of our planet that made all life forms, including human life, possible and sustainable. Punctuated by volatility, with massive tectonic movements, the planet became a complex arrangement of continents. According to many scientists, changes in the natural world provided the conditions for evolution. As world historian Fred Spier has noted, "plate tectonics may have played a dominant part in driving biological evolution, including human evolution. Continuous shifts in the position of the earth's land masses led to changes in the ocean currents, which influenced the global climate. I see this as an example of how dominant geological and climate regimes can influence human evolution."[6]

Although the science of human genetics and evolutionary biology are rapidly changing disciplines, their ability to map genetic and biological structures offers insights into our evolutionary history. In response to geological and climate changes a million and a half years ago, *Homo ergaster*, an early hominid, developed anatomical features more closely identifiable as modern. An increased brain size, a flatter face and jaw, a tilted pelvis for walking upright, and longer limbs distinguished *ergaster* from its hominid predecessors.[7] Cultural changes accompanied the physiological ones. Paleo-anthropologists now believe that the species known as *Homo erectus* and *Homo ergaster* were the same.

3

Our understanding of the co-evolutionary processes of the Earth and of humanity enhances our growing commitment to a historical interpretation based on ecological interdependencies.

Mass Migrations and the Rise of Agriculture

A cooling climate changed the habitats of hominids and precipitated their migration from East Africa to Eurasia, a million or more years ago. With intervals of hundreds of thousands of years separating them, other migrations followed, including the migration of *Homo sapiens* out of Africa about 100,000 BP. Warming climates, retreating ice sheets, the loss of habitat for Ice Age megafauna, and extreme predation by hunters and gatherers may have created a nutritional bottleneck.

The independent invention of agriculture in many parts of the world between 10,000 and 8000 BP may have been a response to these growing nutritional deficits. Rising carbon dioxide (CO_2) levels, an outcome of retreating Ice Age glaciers, made for a more robust plant life that early farmers cultivated selectively. In the opinion of Fred Spier, "the prime candidate for a global factor involved in triggering the emergence of the agricultural regime is climate change."[8] In ecological terms, humans and animals became dependent on each other and on a comparatively small number of domesticated plants.

This transformation from hunting and gathering to agriculture can be described best as evolutionary, occurring in response to ecological changes beyond the control of *Homo sapiens*. The costs to these early cultivators cannot be minimized, however. For millennia, agriculture demanded labor-intensive commitments from humans and their work animals. Their diminished stature and shortened lives suggest a bleak existence.

Adversity led to innovations and inventions, however. The selective breeding of plants and animals and the fabrication of tools increased productivity and led to larger settlements and villages. Many became cities with complex social, economic, and administrative hierarchies. Food surpluses released others to develop skills that increased their mobility and led to a rise in skilled crafts and monetary exchange. With a rise in human skilled capital, many cities became centers for growth and made the transition to agrarian civilizations.

The successful migration of early Eurasian farmers and their livestock spread disease among previously isolated human communities as domestication brought humans and animals into close proximity and exposed humans to animal microbes that became human pathogens. Despite increased food production, a demographic explosion failed to materialize until the twentieth century. A drop in the death rate worldwide, not a rise in the birth rate, caused global populations to skyrocket. The surge did not happen because people "suddenly started breeding like rabbits: it is just that they stopped dying like flies."[9]

Early agrarian societies became a precondition to population growth, manufacturing, and industrialization. As anthropologist Alf Hornborg points out, "land improvement for the purpose of agricultural production is the main form of capital accumulation in preindustrial societies on all continents, and simultaneously one of the most tangible ways in which humans for millennia have changed their natural environments."[10] As this book points out, population growth and the ascendancy of cities became significant outgrowths of the transition to agriculture.

Population Growth and the Rise of Cities

For millions of years, infinitesimal growth rates defined the human population. One theory proposes that the eruption of Mt Toba in Sumatra 74,000 years ago, the largest in the Quaternary (the past 2.6 million years), caused a global cooling that killed all but a few thousand members of *Homo sapiens*. The evidence for this demographic collapse is written into our DNA.[11] Although more than 100 billion people have been born in the past 50,000 years, growth remained very slow for most of human history, not reaching 750 million people until nineteenth-century industrialization.[12]

In antiquity, cities became geographic centers for safety and protection, management, and the distribution of goods and services. From these responsibilities, city-states and agrarian civilizations emerged, with population densities reaching nineteenth-century levels in some centers. Uruk, an early Sumerian city, maintained a population density of as many as 60 persons per acre, with a total population estimated at between 20,000 and 30,000 in 5300 BP, a density similar to Paris, France, in the nineteenth century.[13]

Once cities broke through the artificial barriers imposed by ancient and medieval walls, horizontal low-density growth competed with vertical high-density growth for real and perceived resources. In China and Japan, for example, much of the population growth took place in rural areas after 1750, while in Europe growth occurred in mostly settled and densely populated regions between 1750 and 1850. The impact of growing populations and increasing urbanization placed considerable stress on ecosystems, including woodlands, water, and wildlife, as the ecological footprint of humans broadened and deepened.

In the modern world, these historical transformations and the spread of low-density living arrangements led to more energy consumption producing vast quantities of greenhouse gases. Suburbanites purchase 85 percent more gasoline than those who live within five miles of the city center. This imbalance accounts for about six tons of carbon emissions per vehicle each year.[14] Thinking about cities as "great carbon-reduction machines," where "sidewalks are as sexy as hybrids," and in which cars last for 15 years but street grids last a century or more, may become the template for urban revitalization in the developed world and increasing urbanization in the developing world.[15]

Cities and the Rise of Manufacturing and Industry

Population growth and urbanization can be traced to the growth in manufacturing and industry, two transformations that are co-joined in this book. Manufacturing and industrialization have a long evolutionary history, punctuated by production-changing inventions. At their core, however, is the intensified use of fire, a utility discovered by *Homo ergaster* more than a million years before. As crafts and skilled trades became viable occupational categories in ancient cities, small-scale fabricating and manufacturing using open pit fires and furnaces flourished in urban workshops and in many highly decentralized rural households, across Eurasia, from the Mediterranean to India and China.

Until the eighteenth- and nineteenth-century transitions from workshop to factory, the global economy concentrated on polycentric economic interactions. Asian workshops centered primarily in India and China supplied their trading partners across the world with up to 80 percent of their consumer goods in dyed cotton cloth, silks, and porcelain. Spices, especially pepper from Asia, seasoned the otherwise bland meals across Eurasia. The direction of these energy flows and interactions changed as Europe broke through the bottleneck created by dwindling supplies of organic energy provided by its forests. The coal revolution, extracting mineral energy by burning fossil minerals and converting water into steam, increased production. Steam engines increased economic efficiencies and separated the pre-modern from the modern world. As historian Harold Livesay has written, "the world of material possibilities was dramatically altered between 1780 and 1880. No previous century witnessed such changes."[16]

Industrialization replaced workshops with factories, mercantile exchange with moneyed capitalism, hand looms with power looms, and handicrafts with mass production. Steam turbines replaced steam engines in many industrial enterprises. Inventions, investments, and innovations became synonymous with wealth creation, while toiling masses flocked to cities in search of a better life. As the distance between the classes grew greater, with wealth concentrated in the hands of entrepreneurs and the owners of production, poverty became widespread.

Economic wealth creation imposed heavy environmental costs as industrial cities experienced explosive population growth during the nineteenth century that overwhelmed their ancient infrastructures. Deafening noise, the unbearable smell of rotting draft-animal carcasses, the bloody remains of slaughtered cows and pigs running in the gutters and alley-ways, and human waste from overflowing cesspools, cellars, and privy vaults were a constant reminder of urban decay.

In the *longue durée*, these same wealth-creating industrial cities would become engines of progress. Pockets of poverty would remain, and income would continue to be distributed unequally, but in comparison to rural areas, an extended life expectancy would become a characteristic of living in modern twentieth-century cities. As industrialized cities became engines for the production of capital goods (e.g., iron and steel, locomotives and freight cars, cloth and finished textile goods), they also became vehicles for the growth and spread of consumer goods.

World Trade and New World Ecology

As was the case with manufacturing and industry, consuming material goods cuts a broad path across five centuries of product initiation, marketing, and replacement. Unlike industrial work, however, with its eighteenth-century origins, the patterns of commerce in consumer goods across Eurasia and among Indian Ocean trading partners extend across many centuries. The history of consumer goods, with the transformation of luxuries into commodities for large numbers of buyers, provides readers with a historical perspective into the modern world of mass consumption and its ecological effects. Its location in this book is the logical outcome of global environmental transformations in agriculture, population change, and urbanization.

The international market of exchange and its impact on the global flows of energy, people, and goods became unprecedented. The increased production of sugar and its declining price were tied directly to slave labor. Importing West African slaves to the Americas for work on sugarcane plantations not only transformed sugar consumption in the western world, it also altered human relationships for centuries into dominant and subservient categories.[17] The Sun supplied the energy that transformed seeds into sugarcane and tea plants, while ships catching the trade winds brought slaves from West Africa to the West Indies and Brazil and railroad engines burning fossil coal brought both tea and sugar to consumers.[18] J. R. McNeill revealed how ecological interdependencies shaped geopolitical history in the American tropics. As slaves arrived in the New World in the seventeenth century to grow sugar, the slave ships brought with them the West African mosquito *Aedes aegypti*, vector for yellow fever. Many West Africans had acquired immunity, while Europeans coming from temperate climates had not. Transformed plantation landscapes in the American tropics became breeding grounds for *A. aegypti*. Invading armies from France and England suffered greatly from outbreaks of yellow fever as they tried to defeat the Spanish in the Caribbean.[19]

Fossil Fuels and Climate Change

Energy is a ubiquitous category for which the reader can find evidence throughout this book. While the burning of fossil coal released humans from the constraints of pre-modern organic sources of energy (e.g., wood, biomass, and animal dung), its use began the long-term process of environmental degradation and destruction by emitting carbon, sulfur, and nitrogen oxides that degraded air quality and caused the atmosphere to warm at an alarming rate. The transition to fossil coal accelerated industrialization, after centuries in which humans had harnessed the velocity of the wind and the natural flow of water to do work. Their energy, supplemented by human labor, powered the looms of textile mills, the saws and lathes of lumber mills, and the crushers, cutters, and grinders of ironworks. Most of these power sources and their industrial output trod lightly on the land and its water, when compared to coal-fired plants.

Coal, cheaper than wood, changed the workplace and household economy in the nineteenth century and with it the natural and built environment. Coal continues to fuel industry and generate electricity for modern households. Taking on some of the present demand for energy, petroleum, natural gas, and, to a lesser extent, nuclear and bio-fuels fill increasing worldwide demands. With automobiles becoming the ubiquitous symbols of status throughout the world, petroleum-producing countries play an increasingly significant role in the global economy. With rising demand for energy resources, global greenhouse emissions from all fossil fuels contribute to rising temperatures and climate change.

Since this book began with the many ways in which geological changes created a climate system that made possible the origins of life, a concluding chapter on the world's warming climate serves to reinforce natural world/human history interdependencies. The 2007 report of the United Nation's Intergovernmental Panel on Climate Change (IPCC) serves this purpose, arguing persuasively that the concentrations of carbon dioxide (CO_2) in the atmosphere in 2005 exceeded the natural range of 180 to 300 parts per million found during the past 650,000 years. With an ongoing gathering of information, including 2007–8 findings noting that industrial emissions from China had risen beyond IPCC projections, their forecasts may have understated the hazards posed by a continuing warming of the global climate.

CHAPTER ONE

AN EVOLVING EARTH

Introduction

The Earth is a living, dynamic, and sometimes violent planet. Able to sustain organic life, subject to violent swings in temperature, the modern climate is the result of Earth-defining changes. These changes range from the collision of the Earth's continents to the ability of its oceans to absorb solar energy as heat and reflect it as light, and to influence the intensity and flow of ocean currents. Atmospheric changes, tectonic movements, and global climate are interconnected forces that transformed the Earth's history and created an environment suitable for the development of all life forms.

On a much longer timescale, scientists believe that our solar system formed because gaseous clouds of debris from older stars condensed into solid matter about 4.6 billion years ago (bya). The formation of our solar system was but one of many significant cosmic events in the history of the universe that most scientists agree took place about 13 billion years ago with a Big Bang. At many trillions of degrees, a super-heated universe, smaller than the size of an atom, began expanding faster than the speed of light in its first few seconds. This was the Big Bang. The universe's background radiation is a reminder of this event and the continuing expansion of the cosmos.

For the next 300,000 years, the universe remained a super-heated entity much like the interior of the Sun in our solar system. As the universe expanded and cooled, a phenomenon that continues to this day, energy and matter separated. As described by the historian David Christian: "About 300,000 years after the big bang, all the ingredients of creation were present: time, space, energy, and the basic particles of the material universe, now mostly organized into atoms of hydrogen and helium. Since that time, nothing has really changed. The same energy and the same matter have continued to exist. All that has happened is that for the next 13 billion years these same ingredients have arranged themselves in different patterns, which constantly form and dissipate."[1]

By collapsing the timescales of the 13-billion-year cosmos by a factor of one billion, the Big Bang took place 13 years ago. In this scenario, the Earth's first living organisms appeared about four years ago; modern humans evolved in Africa about 50 minutes ago! The invention of agriculture and the building of cities, which you will read about in later chapters, occurred five minutes and three minutes ago, respectively.[2] Thought about in this way, the life of humans is a relatively recent addition to Earth history. Looked at in another way, if the 10-billion-year projected life of Earth's energy system was compressed into a single year, then all of written human history would be represented in less than a minute. And the twentieth century would be less than a third of a second long.

The Origins of the Earth: From Hot to Cold Planet

More than 4.6 billion years ago the explosive atomic energy of mega-sized meteors created a liquid mass of molten rock, 1800 °F. During a 600-million-year period, repeated bombardments followed by the sinking of the iron cores of those meteors created the molten center of the Earth. The Earth's iron core created the planet's magnetic field that deflected many high-energy and dangerous particles from the Earth. In this very important way, it acquired and to this day possesses a protective shield that allowed delicate life forms to evolve.[3]

This extremely hot planet created an equally torrid atmosphere. The young Earth can be thought of as a massive volcanic field.[4] As large meteor strikes slowed over a period of about two billion years, however, the surface and its atmosphere changed significantly – from hot to cold. Some scientists attribute this climatic transition to the impact of a mega-meteor, estimated to be the size of the planet Mars, that struck Earth about two billion years ago. The impact knocked it on its side, deflecting much of the sunlight that would have normally warmed the tropics. A decrease in solar radiation cooled the surface of the Earth, "dried" the atmosphere causing more radiation to escape from the

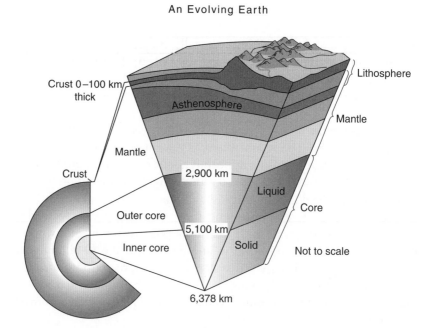

Figure 1.1 Cutaway views showing the internal structure of the Earth. *Source*: Jacquelyne Kious and Robert I. Tilling, *This Dynamic Earth: The Story of Plate Tectonics* (Washington, DC: US Geological Survey, 1996).

surface, and initiated glacial expansion and a colder, snow-covered planet (2.3 bya). The advancing glaciers reflected abnormal amounts of light and heat back into space allowing for further cooling.

A cooling atmosphere made conditions suitable for ice comets emerging from the deep voids in our galaxy and weighing between 20 and 40 tons to bombard our inner space every two to three seconds. According to this "Snowball Earth" theory, these galactic events saturated the Earth's atmosphere with increasing amounts of condensation. "Cosmic rain" cooled the white-hot planet and created the world's earliest oceans. Although the cooling did not require oxygen, free "atmospheric oxygen levels probably increased considerably about 2 bya and again near 800 mya, coincident with major evolutionary changes in Earth's biosphere. Carbon dioxide (CO_2) levels are also believed to have been substantially different during the Precambrian Epoch (4.6Bya–570Mya)."[5]

Increases in CO_2 levels with an atmospheric concentration of 0.035 percent after the Precambrian affected the Earth's atmosphere.[6] Changes in the land and sea biospheres transformed carbon dioxide into oxygen (O_2) and sequestered carbon in various "sinks" – the oceans, mountain ranges, and solid rocks. Accordingly, the planet experienced glaciers that covered the land and water from the northern hemisphere to the tropics about 700 million years ago (mya). The ice sheets locked up about 25 percent of the Earth's carbon. With ice-covered rather than liquid oceans, the oceans could not capture CO_2, released by volcanic eruptions. Under normal conditions, liquid oceans would absorb carbon dioxide. Under these circumstances, however, the atmosphere captured increasing amounts of this soluble gas.

Incidentally, the Snowball Earth theory provides a plausible explanation for the coming of the ice but not an answer to the vital geophysical problem about how the Earth eventually corrected itself to its current $23.5°$ tilt. To date, the most plausible scientific explanation is that the Earth's land mass was clustered together at the South Pole 600 million years ago. The weight of this clustering tipped the Earth into its present inclination. Eventually this land mass broke up to form the continents, a topic described in greater detail later in this chapter.

Icehouse Planet/Greenhouse Planet

These geological events took place over millions of years during the Precambrian when the Earth was ice-covered. However, increasing levels of CO_2, a heat-trapping gas, triggered global warming. When that happened, millions of years of accumulated global ice melted away. As atmospheric temperatures rose, evaporation from oceans and surface water increased, and the Earth's climate progressively became warmer. Water vapor is the largest natural greenhouse gas because it traps solar radiation emitted from the Earth's surface.

In the beginning, these processes created the near-perfect environment for early life forms, including oxygen-producing blue-green algae. Rising levels of oxygen in the atmosphere "made possible the more complex chemistry used by multi-celled animals."[7] The original greenhouse effect triggered the "Cambrian explosion" (570–530 mya), the burst of life that saw the origin of many species.

As current debates about the potentially harmful effects of global warming on the climate system continue, billions of years ago, global warming complemented the production of oxygen-producing blue-green algae, triggered ice melt, and promoted the reaction of oxygen and iron in water.

The presence of oxygen caused iron to separate from water and form deposits at the bottom of the oceans. Additionally, extremely high surface temperatures caused considerable evaporation and consequently considerable rainfall. "Rainfall would have washed the carbon dioxide out of the air, turning it into carbonic acid. The acids then weathered continents, and through chemical reactions, those sediments turned into limestone when they were washed into the oceans."[8] Iron and limestone deposits collected in ocean-bottom sediments that in geological time became deposits in future glacial rocks and were very important to future human manufacturing and industrial activities described in later chapters.

About 550 million years ago, changes in our planet's global carbon budget resulted in significant changes in the global climate. During this entire period, the amount of atmospheric CO_2 dropped to about two or three times modern levels. These declines occurred as carbon became locked in solid rock formations of all kinds, as the ocean floors spread out, and as land-based ecosystems with their diverse and expanding flora evolved.

While this warming continued to green the planet, decreasing amounts of CO_2 caused a slow and gradual cooling of the Earth over millions of years. These fluctuations in CO_2 levels also provide a plausible explanation for the cyclical history of "ice ages" experienced during the past 150 million years. A seemingly regular pattern of ice ages has appeared more recently during the past 740,000 years. Every 100,000 years an ice age of approximately 95,800 years is followed by an interglacial period of approximately 10,000–30,000 years. Given this pattern, human civilization has blossomed during the current interglacial.

Plate Tectonics, Super-Continents, and Climate Change

In addition to the volatile changes in the global climate system, major alterations in the topography of the Earth and in the location of the continents accelerated widespread cooling over millions of years. The movement of major plates, including the breakup of major landmasses into the continents that we know today, changed the climate in ways that altered the ecology of the planet.

As the Earth's surface cooled during its four-billion-year history, the 60-mile-thick crust called the lithosphere was created. In time, volcanic activity and earthquakes caused the lithosphere to break into huge tectonic plates. Repeated activity at the boundaries of these plates caused them to move and drift. Lava flows continued to create and reconfigure the fabric of the Earth's surface. Giant cracks in the plates permitted magma to flow from the Earth's epicenter to its surface. Old rock formations fractured as magma and molten lava flows created new surfaces and new sea floor. The explosive capacity and intensity of these disruptions caused some oceans to expand and others to contract. Mountain ranges pressed upward as giant earthen plates crashed against each other.

The process that explains the movement of these plates is called continental drift and its origin emerged from the hypothesis that in the distant past a super-continent existed. Eventually, it broke apart into several pieces that drifted apart over millions of years of Earth history. Stitching the present-day continents together, as you would a puzzle, along their continental shelves and not along their shorelines, demonstrates the original configuration of this super-continent. That drift occurs seemed plausible but explaining scientifically the source of the energy that powered the movement remained elusive until the second half of the twentieth century.

Then, the development of sonar technology allowed scientists to map the ocean floor. The motion of the plates moves the floor of the ocean into deep trenches where the plates and sediments are carried into the Earth's intensely hot interior. There, they melt and return as lava to the surface of the sea floor. Called subduction, this cyclical process of submerging old sediment reduced the size of the sea floor and simultaneously replaced it with molten lava. In the process, it spread the size of the ocean floor. "In other words, it is the heat of the Earth's interior that provides the power needed to move great plates. That heat is generated largely by radioactive materials within the Earth, which had been formed in the supernova explosion that occurred just before the creation of our solar system."[9]

Although rigid, the Earth's eight large tectonic plates and seven small ones are not static but rather part of a dynamic process of development and change in the material composition of the Earth. Over millions of years, the world experienced a complex series of massive disruptive volcanic activities and earthquakes followed by periods of tranquility. These collisions and uplifts had widespread effects. Throughout geological history, these tectonic plates fused into super-continents, and

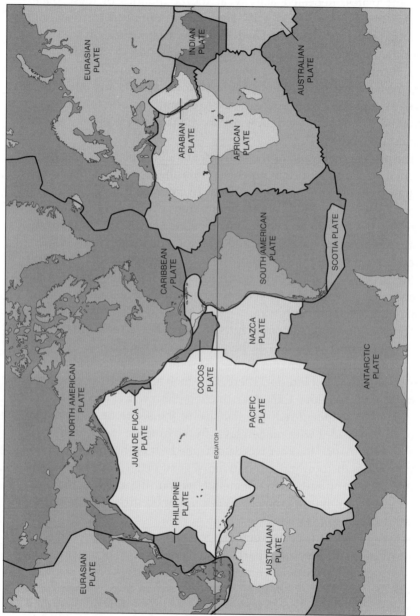

Figure 1.2 A dozen or so rigid slabs called tectonic plates by geologists. *Source:* Jacquelyne Kious and Robert I. Tilling, *This Dynamic Earth: The Story of Plate Tectonics* (Washington, DC: US Geological Survey, 1996).

their eventual breakup and continental drift created the present continents. Their constant movement also had a powerful effect on global climate change that depended on the positioning of major land-masses over the north and south poles.

About 280 million years ago, the continents in the southern hemisphere, namely Antarctica, Australia, Africa, and South America and the landmass that we now know as India, were one large continent known as Gondwana. The existence of major glaciers may have coincided with low CO_2 concentrations. North America, Europe, and Asia were separate floating continents along with glacial ice on the world's ocean. Over 10 million years, Gondwana and the separate floating continents drifted together to form the super-continent, Pangaea (meaning "all Earth"). "Whether we describe it as a landlocked planet with an immense saltwater lake, or an ocean planet with an immense island is only a matter of definition. It might have seemed a friendly world. At least, you could walk anywhere; there were no distant lands across the sea."[10]

The Warming

Pangaea came into existence about 270 million years ago, a time of dramatic climatic changes on the Earth. After millions of years of major glaciations, global climates began warming again. The dew point rose quickly and the ice planet melted and became covered with great swamps.

Before its breakup, Pangaea was composed of large connected land-masses that we now "see" as separate entities. Greenland and the British Isles were a part of Europe. Indonesia, Malaysia, and Japan were connected to Asia. Siberia and Alaska were one, and although now extinct, large, shallow, inland seas covered much of today's landscape. Within 15 million years, however, shocking changes again visited the planet as Pangaea began to fragment. A major volcanic eruption of super-plume magnitude ejected molten lava from the Earth's core, through its crust to the surface. "Texas, Florida, and England were then at the equator. North and South China, in separate pieces, Indochina and Malaya together, and fragments of what would later be Siberia were all large islands. Ice ages flickered on and off every 2.5 million years, and the level of the seas correspondingly fell and rose."[11]

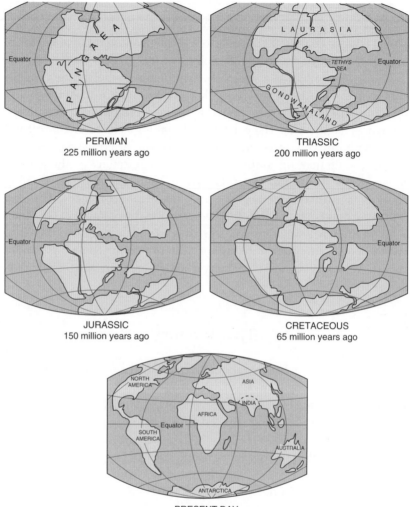

Figure 1.3 Breakup of Pangaea. *Source*: Jacquelyne Kious and Robert I. Tilling, *This Dynamic Earth: The Story of Plate Tectonics* (Washington, DC: US Geological Survey, 1996).

The Earth's surface had been entirely transformed by the end of this 15-million-year period. Lava flows buried entire regions. They reworked Siberia's landscape. What remained of Pangaea drifted northward, moving Siberia to its present location, closer to the North Pole.

"Mega-monsoons, torrential seasonal rains on a much larger scale than humans have ever witnessed, drenched and flooded the land. South China slowly crumpled into Asia. Many volcanoes blew their tops together, belching sulfuric acid into the stratosphere. The biological consequences were profound – a worldwide orgy of dying, on the land and at sea, the likes of which has never been seen before or since."[12] These tectonic fractures and volcanic eruptions caused elevated global levels of CO_2, pushing the climate system into the extended warmth of the Mesozoic Era (230–65 mya). Increased tectonic activity, volcanism, and the weathering of carbon-rich rocks and sediments in geological time elevated CO_2 levels, a major cause of past climate change.[13]

As the breakup of Pangaea continued, the Earth as it looks today began to take shape. By 100 million years ago, a narrow ocean strait separated the two large pieces of the super-continent's puzzle, now known as South America and Africa. They would slowly continue to move apart as sea floor spreading along the enormous mid-ocean rift expanded the Atlantic Ocean's size. About 15½ inches of ocean floor was created every one million years. Simultaneously, older ocean floor disappeared into deep ocean trenches or forced itself under continental land masses, the subduction zones described earlier. This drifting and colliding caused considerable tectonic instability. Today, the Atlantic Ocean grows about one and a half inches a year or at about the same rate as our fingernails grow.[14] Throughout this lengthy geological history, the Earth was mostly ice-free, suggesting the existence not only of a warm climate but of elevated CO_2 levels as well.[15]

The Cooling

Further fragmentation of the remaining super-continent, Gondwana, in the southern hemisphere triggered the transition from a global greenhouse to a cooler planet. The release of trapped frigid waters at the South Pole, continental drift and collision, and the creation of the Tibetan Plateau, as well as many other significant geological events, accelerated this transition. India separated from Antarctica 130 million years ago and began its slow drift toward the Asian subcontinent. About 60 million years ago, what is now the island of Mauritius and the continent of Australia began to separate from Antarctica. By 33 million years ago, the final underwater land connection between Antarctica and Australia and stretching south of Tasmania was broken, creating *la grande coupure*, the great cut.[16]

As Australia drifted northward, an ocean with deep cold bottom waters passed through the great cut, creating the thermal isolation of Antarctica. During the earlier "greenhouse" phase, Antarctica had supported abundant temperate-climate vegetation. Today, 8,155 cubic feet of frigid ocean water pass between Antarctica and Australia every second. This volume represents one thousand times the flow of the mighty Amazon River every second.[17] Called the Antarctic Circumpolar Current by scientists, it has been identified as the condition that ended Earth's lengthy greenhouse period and initiated global cooling and its cycle of ice ages. By keeping cold waters circulating around Antarctica, it created the southern polar ice cap and maintained the cold ocean bottoms around the Earth. According to Australian ecologist Tim Flannery, "the establishment of the Antarctic Circumpolar Current was the switch that turned on the modern world ... at the time of the great cut mean global temperatures dropped by an extraordinary five to six degrees to just 5 degrees Celsius (41 °F)."[18]

In the southern hemisphere, icebergs from Antarctica began appearing more than 20 million years ago, about 35 million years after the separation of Australia from Antarctica. Caused once again by ocean floor spreading, the separation "made possible the free flow of oceanic currents around the South Polar continent, the effect, for a time, was greatly increased snowfall on the South Pole area, producing the heavy flow of ice that occurred five million years ago."[19] The ice spread far beyond the continental boundaries, creating ice cliffs almost 1,000 feet high. As they grew in size and weight, they descended into the ocean depths as much as 1,900 feet, scouring the ocean floor and creating deep cold-water ocean bottoms. At no time in human history has this region been free of ice.

In North America, mean temperatures dropped more than 18 °F, resulting in massive extinctions of marine life, flora, and fauna and the onset of seasonal variability with tropical summers and frigid winters. Before this cooling, rainfall covered the world more evenly throughout the year than today. Without seasonal dry periods, grasslands, varied vegetation, and deserts were uncommon. Covered mostly by oceans and shallow seas, global temperatures stabilized worldwide. Arctic seas, frozen regions, and glaciers were either nonexistent or severely limited. With global cooling, mixed evergreen forests gave way to mixed hardwood forests in North America. Changes occurred throughout the world as the modern glaciated world came into existence.

As the cool climate replaced the warm and the wet by the dry, shallow coastal and inland seas that covered much of North America retreated, uncovering larger bodies of land. With cooling, temperatures worldwide became more variable. As a consequence of this cooling, a reduction in precipitation allowed open habitats of savanna and grasslands to replace dense forests. Although the long trend was clearly in the direction of a cooling of the Earth and the eventual formation of glacial ice, climatic fluctuations from warmer to colder and vice versa continued. These climatic fluctuations with the newer phases of cooling and drying led to the development of summer and winter. "Seasonality requires plants to adapt once- or twice-yearly reproductive phases, with seeds that can survive bad weather while lying in a dormant state. The grasses were particularly successful at doing this, resulting in open savanna grasslands."[20] Happening slowly, this transformation proved to be a significant change leading to the development and evolution of mammal life. These geological events exercised a controlling influence on faunal evolution and were essential to the eventual evolution of hominids, the ancestors of anatomically modern humans.

The Elevation of the Tibetan Plateau and Its Effect on the Global Climate

During a 10-million-year period from 66.4 to 57.8 mya, the Earth experienced increasing continental uplift. India collided with Asia, and "the compression stresses that built up after the two continents collided somehow forced Tibet upward, although the exact mechanism responsible for the uplift is still being debated."[21] At the point of collision between India and Asia, rock from both was pushed, compressed, crushed, and pulverized. Some was pushed upward to create the Himalayas, the youngest mountain range in the world. Some was pushed downward into the mantle of the Earth to create the base of the mountains in order to prevent them from collapsing under the weight of the uplift. The density of the mantle thickened. The mountains themselves represent this thickening of the Earth's crust brought about by upward compression.

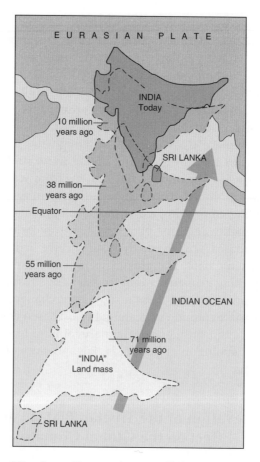

Figure 1.4 The Himalayas: Two continents collide. *Source*: Jacquelyne Kious and Robert I. Tilling, *This Dynamic Earth: The Story of Plate Tectonics* (Washington, DC: US Geological Survey, 1996).

As Asia and India collided, the ocean plate and part of the landmass of India pushed under the edge of Asia. Some of India's land was not as dense as the Earth's thick mantle, however, so rather than being pushed downward, it moved horizontally, creating a wedge between Asia and India. The impact of this horizontal movement uplifted the Tibetan Plateau as one solid piece of rock. "The rock called (the continental crust) beneath Tibet is an astonishing 46 miles thick, about twice the average crustal thickness (of the Earth)."[22]

Called the "roof of the world," the Tibetan Plateau is almost a third as large as the continental United States. At 15,000 feet in elevation, or three miles above sea level, it exceeds the height of most of the Rocky Mountains. The Himalaya Mountains, seven miles above sea level, form a 1,800-mile-long rampart between India and Central Asia. For geologists and climatologists, how this uplift changed global climates and created the conditions for the evolution of humans remains an important question.

As the formation of the Tibetan Plateau and the Himalayas coincided with the continuing transition from warm worldwide "greenhouse" climates to more temperate climates about 24 million years ago, the evolutionary pattern for warm-blooded species became sustainable.

Seasonality occurred because the building of the Himalayas and the Tibetan Plateau changed the climate of Asia. In Asia, seasonality, with its dry climate during the winter and its rain in the summer, created the monsoon. "This occurs because the vast, high Himalayan region is quickly warmed by the summer sun, heating the air and causing it to rise. This draws in moist air from the sea and the moisture falls as rain."[23] Before these climatic changes began, Tibet possessed tropical and subtropical forests. As recently as 10 million years ago, the region contained deciduous forests very much like those found in the temperate regions of the world. Today, the vegetation differs markedly from that of the past. It is primarily grass and scrub vegetation, "adapted to the harsh steppe climate, which involves severe winters and seasonal drying."[24]

For about every change in elevation of approximately one and a half miles, the temperature drops about 32 °F. The changes in vegetation and the evidence from sedimentary formations suggest that the Tibetan plateau has risen about three miles above sea level in the past 25 million years, with one-half of the uplift occurring during the past 10 million years. The adjacent regions of southern China and Southeast Asia, unaffected by the uplift, grow vegetation suited for a warm climate; one that the fossil and pollen remains in Tibet today suggest was present there 40 million years before the uplift. "Even more definitively, uplift accounts for the presence of ocean sediment ... approximately 70 million years old in Tibet."[25]

What was the effect of the Tibetan Plateau uplift on climate change in the Eurasian regions affected by the tectonic collisions of India into

Asia? First, the uplifted region blocked the natural west-to-east wind currents at the surface and in the upper atmosphere through the middle latitudes of the Earth. With the westerly winds blocked, the east winds approaching the plateau were diverted northward around the plateau. One additional result was a "meander" or large southward return flow. These changes in circulation represent long-term climatic patterns rather than short-term weather reporting.[26]

Second, a high plateau affected winter and summer air circulation. Without the uplift, precipitation stretched out over a larger geographical area and fell with less intensity. The sun rapidly heated the thin atmosphere of the uplifted areas. Because of its low density, warm air rose rapidly. Rising air cooled and lost its capacity to hold water vapor. Since moisture from the Indian Ocean was blocked by the lower atmospheric pressure of the plateau, more intense rainfall in the form of summer monsoons hit the southeast region.[27] India and Southeast Asia became warmer and wetter as a result of the Tibetan Plateau. In their wake, monsoons today bring flooding to the low-lying areas of the Indian subcontinent, soil erosion, the destruction of crops, and the loss of life.

Third, the effects of seasonal heating and cooling were global. "In the summer the rising motion of air over the plateau leads to a compensating sinking of air over surrounding regions, including the high-pressure regions that lie over the oceans (which are cool compared with land temperatures) at subtropical latitudes. Heating over the huge Tibetan Plateau also induces air to sink over adjacent areas in the Mediterranean and central Asia. The sinking air is dry because it originates from high elevations far from oceanic moisture sources. Compressional heating, which takes place as the air sinks, also lowers the relative humidity."[28] The reverse occurred during the winter months. Air over the plateau fell as air over the low-pressure oceans rose. Warmer ocean temperatures prevailed relative to the cold air masses of the elevated plateau. Like the summer, the effect of falling air pressure on the plateau in the winter had global climate effects that altered the climates of Asia, Africa, and Europe.

The summer winds that contained more moisture switched from the west to a drier northeast. The drier air of the elevated region blocked moisture coming from the Indian Ocean. These climatic conditions explain the existence of cooler summers and colder winters in northern Asia. During the millions of years of continental uplift, the flora changed to reflect the gradual changes in climate. Over time the sub-arctic forests and tundra replaced the warm deciduous forests of 20 million years ago. The gradual transformation is shown by the evolution "... from forest to steppe and even to desert vegetation, as shown in the fossil record of the past 20 million years."[29]

Because of the uplift, Europe's atmospheric circulation wind currents shifted in response to the location and size of the Tibetan Plateau. Since wind arrived from the northeast and flowed in a north to south direction, the continent experienced colder and drier winters and cooler summers. Because of these wind currents, the Mediterranean, the Arabian Peninsula, and northwest Africa were drier. The evidence to support these and other observations was found in sediment. Deep-sea sediments contain millions of years of dust and debris blown across the continents into the oceans of the world. During periods of continental aridity, larger amounts of wind-blown sediments appear in the oceans, serving as indicators of atmospheric wind intensity and direction.

For example, the sediments, calcium carbonate, wind-blown quartz and clay, and siliceous microfossils found in the Arabian Sea, blown from the arid African continent, serve as useful measures of the extent of the aridity and the strength of the summer Indian Ocean monsoon. Also, sediments found in the North Pacific Ocean east of Japan were the remains of wind-blown soil from central China 2,400 miles away deposited during five separate 95,800-year glacial cycles.[30]

So, the collision of India and Asia and the continuing pressure of plate movement had profound ecological and climatic effects in the region and beyond. The uplifting of the land into broad plateaus and the building of mountain chains changed the Earth's climate. Continental mountain building resulted in cooler temperatures in the areas of the uplift, and because of the altered atmospheric circulation patterns had global cooling and drying effects.

For example, early climate records of the Bighorn basin of Utah and the Badlands regions of South Dakota in the American West support the conclusion that the onset of aridity was triggered by continental uplift. Both experienced an average annual rainfall of 40 inches before the uplift and less than 16 inches thereafter.[31] What geologists had been claiming for more than a century, namely that mountain building can cause climate change, has been essentially proven by climatologists. "One potentially important factor was the uplift of the Tibetan and North American plateaus, which have led to elevated continental weathering rates and the drawdown of atmospheric CO_2."[32]

High mountains and plateaus contrasted strongly with the prevailing low continental landscapes during the earlier greenhouse phase, millions of years before. "The evolution of certain plant types about seven million years ago that were capable of a distinct photosynthetic pathway for carbon (called C_4 plants) may have resulted from lower atmospheric CO_2 levels."[33] Since both the Tibetan and North American plateaus rose substantially during the past 40 million years, they provide a plausible explanation for global cooling.

An abrupt warming, by several degrees, punctuated the Earth's tectonic activity. It occurred approximately 55 million years ago and lasted for about 100,000 years. Its environmental impact was felt in the northern and southern hemispheres and in the deep oceans. Because this approximately 100,000-year anomaly occurred at the Paleocene and Eocene boundaries, it is often referred to as the Paleocene–Eocene Thermal Maximum. Shifts in global rainfall patterns and in vegetation are present in various fossil records. Because this warming anomaly took place during a major reshaping of the Earth's landscape and an overall cooling phase, the relationship of these events remains a subject for scientific inquiry.

Fossilized ocean and continental records reveal a release of massive amounts of carbon from any of these likely sources: volcanic activity during tectonic uplift, the discharge of methane (CH_4) from the decomposition of clathrates on the ocean floor, or the oxidation of organic matter.[34] This event has become one that is studied closely by climatologists because the rapid release of carbon into the atmosphere by natural causes is similar to the release today by humans. The effects of the former not only changed atmospheric conditions, raising temperatures 5–8 °C, but it changed deep ocean circulation patterns in the span of a few thousand years. That it took an additional 100,000 years for the climate system to recover from the Thermal Maximum has not been lost on scientists attempting to predict the length of time that it would take for the world to recover from the warming since the beginning of industrialization.[35] "The event is a striking example of massive carbon release and related extreme climatic warming."[36]

The Birth, Death, and Rebirth of the Mediterranean Sea and its Hemispheric Environmental Effects

The breakup of Pangaea ended 50 million years ago and a new era of continental collisions began. Africa struck Europe, creating the Swiss Alps. The tectonic pressure of Africa moving northward against Arabia and the Eurasian land mass squeezed the ancient Tethys Sea which once separated Africa from Arabia. Once it collided with Eurasia, only a small eastern section of the Tethys Sea remained as a small remnant of the once great warm-water belt that dominated the region from the Atlantic to the Indian Ocean and swamped large areas of Europe, northern Africa, and the middle of Southwest Asia. For much of its 300-million-year existence, the Tethys Sea was inhabited by an extensive variety of warm-water invertebrates living in average water temperatures of 77 °F in areas that today are in the cold temperate zone. Twenty-five million years ago, sand dollars lived everywhere and coral reefs dominated large areas of this sea.[37]

The pressure of folding plates and uplift of the deep Tethys Sea sediments created the mountain ranges from the Alps to the Caucasus in Ukraine and the Zagros in Iran. When the two continents separated ever so slightly, the Mediterranean Sea was all that remained of this ancient sea. Today, the pressure exerted by the movement of the African tectonic plate is responsible for the volcanic and earthquake activity that extends from Portugal across the northern coast of the Mediterranean Sea to Turkey and Iran.[38] Colliding continents caused continental compression and led to wider ocean basins and lower sea levels, thereby creating more land surface. A larger surface permitted easier movement of flora and fauna. More land usually meant more plants to reflect much more solar heat, possibly aiding in the cooling of the planet.

With colliding continents, exchanges of plants and animals from Africa and Eurasia became commonplace. Horses and bovines spread to Africa while elephants and primates dispersed from Africa to Eurasia. The movements of human species (hominids) from Africa represent an important chapter in this global exchange. "This topographical configuration, and the population movements it permitted, was the topological foundation for the emergence and spread of successive

27

species of *Homo.*"[39] This migration of hominids will be discussed more fully in the following chapter.

Ecological changes and their impact on human evolution are reflected in the 20-million-year history of the Mediterranean Sea. Before the closing of the Tethys Sea, cold ocean water flowed freely from what was to become the Atlantic and Indian Oceans. It sustained a diverse marine life from the cold sedimentary bottom to the warmer sub-surface levels. The climate of the surrounding landmass was also moderated by the open passages from one ocean to another. About 15 million years ago, the connection to the Indian Ocean closed. Tectonic pressure caused by converging plates and mountain building closed Gibraltar sometime between eight and five million years ago. The Mediterranean Sea became isolated from the Atlantic Ocean, and in the process began its thousand years of evaporating into a dry desert.

Given the following calculations, it would take about 1,000 years for the Mediterranean Sea to evaporate completely. Taking into account the climate in the region today, about 2,400 cubic miles of seawater evaporate each year. Less than 300 cubic miles of water from rain and freshwater from rivers emptying into the sea replace it each year. Through the narrow Straits of Gibraltar, the sea receives about 2,100 cubic miles of seawater each year to maintain it at a constant level.[40] Despite these calculations, however, it seems that the Mediterranean did not evaporate that quickly eight million years ago because freshwater from rivers continued to empty into it. "… Eastern Europe … was covered at that time by a vast body of brackish water, called Lac Mer. This extended all the way from Vienna to the Ural Mountains and the Aral Sea. The present Caspian Sea and Black Sea are its last remnants … The bottom of the Mediterranean basin [was] partly covered by a series of large, brackish lakes for some time between 8 and 7 million years ago … They disappeared about 7 million years ago, when the converging plates elevated the Carpathian Mountains, sufficiently to change the drainage pattern, and the waters of Lac Mer then escaped to the north … and the lakes evaporated completely."[41]

Without water, continents with exposed shelves surrounded the Mediterranean. High plateaus with steep slopes descended to the floor of the former Tethys Sea. The exposed floor revealed peaks and plateaus of submarine volcanoes. They dotted the former seascape as sunken valleys and basins about two miles below sea level. It was a barren, desert landscape with temperatures reaching 150 °F.[42] The transformation of the Mediterranean into a great desert basin helps to explain the dramatic change in the climate of central Europe from 20 million to 5 million years ago. "The Vienna woods were changed into steppes and palms grew in Switzerland."[43] To reach the desolate and desiccated Mediterranean, the great pre-historic rivers, the Nile and the Rhone, cut deep canyons, at least 900 feet below present sea level. These canyons represent the remains of ancient river systems created when the Mediterranean became a desert.

About 5.5 million years ago, the movement of the tectonic plates that closed the area around Gibraltar's Straits 2.5 million years before opened. The new opening revealed a very high scarp that exposed a steep drop to the Mediterranean Sea floor two miles below. Over time, erosion of the rock barriers that dammed the Straits turned a trickle into a cascade, and finally into a great waterfall. The flow of water from the Atlantic would need to exceed the loss from evaporation by a factor of 10 in order to fill the Mediterranean.[44]

The relationship between land and the flow of water played a major role in the changing climate and environmental conditions of the Mediterranean region. A dry, arid, dusty and inhospitable region would one day become a center for civilization. The role of the rivers that emptied into the Sea and the role of the port cities that bound the region together were inextricably linked to the life-sustaining qualities of the Sea. This relationship also proved to be important in other parts of the world as seafarers from a more recent time would depart from these port cities and begin anew the binding of the world together as they traveled and traded in China, India, and the Americas.

The Impact of the Isthmus of Panama on Global Climate Change

About 10 million years ago, the Isthmus and presumably all of Central America began rising from the ocean floor. This natural damming and separating of the Equatorial Atlantic and Pacific Oceans also altered global climate conditions. On the Caribbean side of the Isthmus, volcanic activity, prevalent from five to seven million years ago, created new land formations from the lava flows. The activity of Pacific sea floor spreading along the Galápagos Islands Rift Zone to the south of the Isthmus provided clues to its rising. The spreading of the sea floor created south-to-north fractures in the entrance of the North American Trench along the west coasts of Costa Rica and Panama.

As new ocean floor emerged along the Galápagos Ridge, the northward movement of old ocean floor descended into the Middle America Trench. By pushing and burrowing its way under the lithosphere, it caused the uplift of Central America, including the Isthmus of Panama. By four million years ago, the closing of the "gate" separating the Pacific and Atlantic Oceans was complete, marking the end of the worldwide equatorial warm-water circular system. With separate circular systems flowing in the two oceans, the Gulf Stream became part of the new Atlantic Thermoline. This change altered the path of storms and delivered warmer, humid air to central Canada and northern Europe. In winter, where heavy snow accumulations exceeded summer-melt over a long period, ice-age conditions prevailed.[45]

The opening and closing of isthmuses contribute to the cooling and drying of the Earth. They separate one large body of water from another and change the ocean currents that alter climate conditions. The closing of the Isthmus of Panama prevented warm Atlantic water from circulating around the equator, accelerating the further cooling of the global climate. It also connected two large continents that prompted the movement of fauna, flora, and much later, humans across this passage. Although humans may have made their way south in boats as well as on foot 30,000 to 15,000 BP, they may have landed on the Isthmus as they traveled south.

The Mid-Pliocene, Glacial and Interglacial Cycles, and "Modern" Times

The Mid-Pliocene (3.3 to 3.0 mya) was a time when mean global temperatures were higher by 2 to 3 °C than those found during the entire pre-industrial period spanning thousands of years, but not before the emergence of hominids in Africa, the earliest human ancestors. By the Mid-Pliocene, the location of the continents and oceans was similar to their current configuration, making global temperature comparisons with those of the modern era possible. Atmospheric CO_2 concentrations were estimated to be between 360 to 400 ppm (parts per million) during the Mid-Pliocene, higher than pre-industrial levels but approaching current levels in 2008. At these CO_2 concentrations, geological and ice core evidence prove that sea levels rose at least 15 to 25 meters above modern levels. (A meter is 39.37 inches.)

An atmospheric-ocean experiment simulating Mid-Pliocene sea surface and air conditions and conducted by scientists in 2005 arrived at the following conclusions. Carbon dioxide concentrations of 400 ppm produced a warming 3 °C to 5 °C in the North Atlantic and 1 °C to 3 °C in the tropics, suggesting that the northern latitudes may be more sensitive to elevated concentrations of CO_2. Among the possible causes were the following: an increased movement of heat from the tropics due to a quicker deep ocean circulation or faster surface winds increased the flow of surface ocean currents.

These possible causes differ from models of twenty-first century warming that postulate a slowing of the North Atlantic Deep Ocean Circulation caused by glacial melt and freshwater discharge into the oceans. Chapter 10, A Warming Climate, will provide a fuller explanation of the relationship between global climate change and deep-water circulations. Clearly, understanding the dynamics of climate warming during the Mid-Pliocene may help in predicting the effects of elevated CO_2 concentrations and the role played by ocean circulations in a globally warmer world.[46] Documentation from ice cores points out that it took 100,000 years for carbon dioxide concentrations to stabilize and return to pre-agricultural levels of 180 ppm.

These same paleo-climate records document the sequence of glacial and interglacial cycles for the past 740,000 years, with evidence from deep ocean sediments indicating other cycles for several million years. The best documentation comes from the past 430,000 years, with glacial/interglacial cycles lasting approximately 95,800 years and the length of the warm interglacial mode for these cycles ranging from about 10,000 to 30,000 years. Within this full cycle, additional cycles occur, responding to changes in the Earth's spin or tilt as it orbits around the Sun. At 41,000 years, one full cycle is determined by the tilt in the Earth's axis that controls the amount of radiation from the Sun in the high latitudes. At either 23,000 or 19,000 years a "wobble" in the Earth rotation causes a much weaker warm/cold cycle with greater effect over the low latitudes near the Equator.

During every 95,800 years during the past 740,000 years, major climate changes resulted from these interglacial/glacial cycles. Forests grew and retreated. Water levels in lakes, rivers, and streams rose and fell and the contours of continental shelves appeared and then were

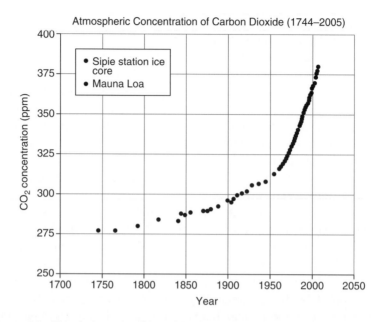

Figure 1.5 Trends in atmospheric CO_2 and global surface temperature – the past 400,000 years. *Source*: Permission granted by Pew Center on Global Climate Change. www.pewclimate.org.

inundated by changing sea levels. In the high latitudes and in high mountain ranges, glaciers advanced. In front of the advancing ice appeared an expanding and broad band of peri-glacial and permafrost landscapes. Climate changes and their impact on the environment were frequent and predicable.[47]

The following brief description of the Earth's "recent" climate history covers more than a million years. At 2.5 million years ago, moderately sized ice sheets existed in the northern hemisphere but were replaced by more extensive glaciers 900,000 years ago. During the Pleistocene Epoch (1.8 mya to 11,600 BP), large ice sheets covered much of the northern hemisphere; this was followed by periods of warming with retreating glaciers. A mostly dry and arid climate existed, with low levels of atmospheric CO_2 suitable for hunting and gathering but inhospitable for plant cultivation or farming. The most recent epoch, the Holocene, covers the past 11,600 years and is characterized by a warming global climate, retreating large glaciers, and growing human populations. During recent decades, scientists have noted the important role of trace gases in abruptly changing the Earth's global environment during the transition 11,600 years ago.

Climate change during the past million or more years and the entire Holocene remains the most intensely studied period in the history of the Earth's climate. Glaciers melted after dominating the landscape and the oceans in the northern hemisphere, covering much of North America, Europe, and the northern reaches of Asia for hundreds of thousands of years. Ice locked up much of the Earth's continental and ocean water, the resource that most directly influences the global climate.

With the melting came warming and an epoch of relative climate stability. Although the long trend was a general warming, the Earth was not uniformly warm. Warm temperatures often exceeded current ones but this interglacial period experienced some decades of significant cooling. Historically, little ice ages, influenced mostly by solar activity, disrupted economic and social life during Europe's first and second millennia. During most of the Holocene, carbon dioxide (CO_2) concentrations remained high, when compared to the glacial, and stable enough to accelerate the transition to plant cultivation and farming. Since 1750, however, burning fossil fuels to power manufacturing and industrial activity have contributed to elevated levels of CO_2 and human-induced climate change.

CHAPTER TWO

EVOLVING HUMANITY

Introduction

"Earth and life evolve together."[1] Geological processes molded the Earth and the Earth's changing geology complemented evolving life forms. The geological changes that created and then dismembered the super-continents reduced and then expanded biological diversity. The creation of Pangaea about 250 million years ago resulted in the extinction of more than 95 percent of the world's marine life. Its eventual breakup promoted a biologically diverse planet of plants, insects, birds, and mammals and eliminated reptilian domination during one of the Earth's major greenhouse phases. Geological processes changed the surface of the earth and in doing so made life forms conducive to speciation. While some species were replaced, others separated, creating new ones. Separation created barriers and it also established the pathways for life to form in the oceans and on the land.

The concept of pathways along major geological breaks and rifts helps to explain how species evolved by genetic replacement. Surviving members of a species adapted to the new environmental conditions caused by geological disruptions. In this way, geology provides clues for understanding the relationship between the location of pathways and barriers and both the existence of some and the absence of other fossil remains in early animal and hominid history. With this understanding, this chapter traces human evolution as it relates to its changing geological and climatic surroundings.[2]

The expansion and contraction of woodlands and tropical forests as a result of ice-age climatic cycles produced isolated and fragmented environments in which populations of mammals, birds, and insects of all kinds lived. Volcanic eruptions and tectonic rifting caused the initially large homogeneous population of plants and mammals to become isolated, separated, and fragmented. Reproducing in isolated environments gradually led to more varied local populations and the evolution of new species. Separated by geological activity into isolated and fragmented communities, they survived by coping with their new environments and changing their behavior. They changed by establishing domains, developing defenses against intruders, migrating to new niches, and passing this information to their offspring.

Geographical isolation ultimately meant that the smaller, separated, and fragmented population possessed a smaller genetic pool. Selection began to work in these less diverse genetic populations. Under such conditions, the opportunistic process of evolution led to rapid divergence through mutation and selection, and to new species.[3] The human species is no exception to this pattern of evolution by natural selection.

Climatic Changes and Evolution

Evolution was intimately connected to tectonic activities and climatic changes. The drier and cooler climate described in Chapter 1 continued for millions of years. It promoted an intense competition among a variety of species for habitat and food that resulted in an accelerated evolution. On these longer timescales, the linkage of climate changes in Africa and the evolution of modern humans represents a new area of inquiry. The disappearance of the Tethys Sea allowed for a variety of animal species to migrate across northeast Africa, Arabia, Turkey, and Iran for the first time. The drying and cooling of the region's climate increased the savanna landmass by shrinking the tropics and promoting the evolution of savanna species including bovids, giraffes, ostrich, modern rhinos, and several primates.[4]

Climatic changes altered the geographical belts of vegetation. Dry, cool air descended on the tropics from glaciers covering the northern latitudes from Eurasia to North America and extending as far south as Britain. It expanded the deserts and caused the tropical forests to retreat and fragment into a complex mosaic of micro-grassland environments supporting a variety of grazing and migrating animals and primates.

Why glacial ice formed in the first place is a subject of ongoing investigation by paleo-climatologists. One theory argues that rain and snowfall increased in northern Europe, caused by a stronger Gulf Stream. Warmer and moist water in geological time had traveled westward from the Atlantic into the Pacific Ocean. With the closing of the Isthmus of Panama, this warm and moist water was diverted northward into the Gulf Stream and diagonally across the Atlantic from the Gulf of Mexico, causing increased snowfall. As a result, glacial ice thousands of miles away in the northern hemisphere and the closing of the Isthmus in the south created a drier climate in Africa.[5]

Recurring ice ages resulted in the expansion of the Antarctic ice sheets and the decline of the global sea levels. As the massive Antarctic ice grew, the ocean water around it cooled. The north-flowing, cold Benguela Current, an extension of the Circumpolar Current described in Chapter 1, became colder as it flowed northward from the Antarctic Sea to the west coast of Africa. Since cold salt water evaporates at a slower rate than warm salt water, the normally high levels of water vapor in the atmosphere decreased, altering rainfall patterns over much of west and central Africa. Precipitation dropped accordingly. Dense tropical forests, normally fed by high levels of precipitation, retreated and extensive savannas replaced them, increasing the competition among existing species for habitat, food, and water. The development of the savannas also helps to explain why Africa became the most suitable location for human evolution through speciation.[6]

The field of human origins and evolution is known for debate and controversy rather than for agreement and consensus. The origins of *Homo* remain one of science's most complex puzzles, with large gaps in our knowledge about its evolution. And it will continue along these paths as paleontologists, archeologists, anthropologists, climatologists, and geneticists continue to unravel the history of humanity. As recently as the 1980s, two scientists, Milford Wolpoff and Alan Thorne, challenged the authenticity of the "Out of Africa" hypothesis that proposed that modern *Homo sapiens* had a single origin in Africa about 200,000 years ago. Wolpoff and Thorne revived the multi-regional hypothesis, arguing for multiple origins using evidence from fossil remains.[7]

In 1987, the geneticists Rebecca Cann and Allan Wilson challenged the multiple origins theory with new scientific findings. In the scientific journal *Nature*, they argued that their population studies of mitochondrial DNA rather than nuclear DNA proved that all modern humans are descended from a common ancestor. Unlike nuclear DNA, mitochondrial DNA is passed only from mother to child, allowing researchers to construct a family tree based on gene structure for all of humanity. Using genetic information in this way suggested two conclusions. First, the gene structure among present-day Africans was more diverse than that of either Asians or Europeans. Second, their molecular clock calibrations suggested that an African woman, living 100,000 years ago or more, was our common ancestor and a member of the genus *Homo sapiens*. In the genetic literature, she has become known as the "mitochondrial Eve."[8] Despite the reference to the Biblical Eve, this mitochondrial Eve was not the first woman to live on Earth. Her lineage, however, is the only one to survive, according to geneticists.

Another Effect of the Closing of the Mediterranean Sea

As the cooling of the global climate continued and more and more ocean water became locked in the expanding Antarctic ice sheets, global sea levels dropped even more, further lowering precipitation.[9] Declining sea levels reduced the flow of Atlantic seawater into the Mediterranean. As noted earlier, the decline continued until the sill at the Straits of Gibraltar became too high for water to enter. The closing had far-reaching evolutionary, environmental, and climatic effects.

Without a ready supply of ocean water, the evaporation of the Mediterranean turned the area into an immense dry salt desert, sequestering about 6 percent of the world's supply of ocean salts. By decreasing the salt content of the oceans, seawater froze more rapidly, expanding the Antarctic ice cap and Arctic glaciers. All of this, of course, caused a rapid fall in global temperatures. Six million years ago and lasting for at least a million years, cool and dry climates expanded desert landscapes and led to the further contraction of the dense and wet African forests and the expansion of dry East African savannas. Hominids found themselves increasingly pushed back and confined to fragmented tropical environments in sub-Saharan Africa. Isolated in this way, the genus *Homo* began the long, slow evolution into separate species.[10]

Global cooling and the changing direction of wind currents caused by continuing tectonic activity brought additional precipitation in the northern hemisphere. Added moisture in the northern latitudes increased the size and movement of ice that cooled and dried the air circulating further south, in some instances as far south as the tropical rainforests. Rainforests continued to retreat, fragment, and become isolated habitats for various species, while the savannas and deserts expanded into new areas.

Rapid change took place at critical junctures during as many as eight glacial/interglacial cycles during the past 740,000 years. Because these cycles appeared at regular multi-thousand-year intervals, changes took place in either expanding or shrinking microenvironments, such as the tropical forests, savannas, and deserts during these intervals. These isolated environments became the cradles of evolution and either grew or contracted because of changes in climate. Since climate change forced relocation for survival, dispersal became an early pattern in the evolution of species, including hominids. Rapid climate variations quickened the pace of evolution. So, periods of significant climate change accelerated the evolution of species and periods of climatic stability reinforced biological stability.

Human Ancestry

If distance between isolated habitats of tropical trees and patches of grassland became a problem in gathering food, then an adaptive mechanism was needed in order to search for food, remain safe from large carnivorous predators, and return to one's habitat. The selective pressure on hominids was to walk, to walk quickly, and eventually to run, and to remain "a step ahead" of even quicker predators. In addition to the selective pressure to cope with this new environment and to hunt and gather, other pressures also help to explain the upright stance and bipedal locomotion among hominids.

Among four-legged mammals, more body surface is exposed to the scorching sun. As a result, they evolved large nuzzles to breathe in large amounts of hot moist air. By panting they increased airflow, causing air to cool and to evaporate. In this way, they regulated their body temperatures and survived in the tropical heat. In addition, they remained in the shade, wherever it could be found, and hunted at night. By standing upright, hominids reduced the area exposed to direct sunlight and cooled faster when exposed to the wind.

The evolution of the brain area called the hypothalamus regulated body temperature. As the temperature of the body rose, the hypothalamus sent impulses through the nerves to stimulate sweat glands and the evaporation of sweat cooled the body. Hair on the head served as a shield against the sun, and in combination with sweat glands promoted brain and body cooling. Each of these evolutionary changes equipped hominids to cope with life in a hostile climate.[11]

Climate also helps to explain variation, inheritance, and selection in the development of physical characteristics during the course of human evolution. Dark skin protects populations living in the tropics from the life-threatening effects of the Sun's ultraviolet rays. Early modern Europeans migrated from Southwest Asia, an environment deficient in vitamin D, to higher latitudes, "… because the essential vitamin can be produced, with the aid of sunlight, from precursor molecules found in cereals. For this, Europeans have developed the whiteness of their skin, which the sun's ultraviolet radiation can penetrate to transform these precursors into vitamin D. It is not without reason that Europeans have, on average, whiter skin the further north they are born."[12]

The Birth of Human Intelligence

Climatic changes and human physical adaptations prompted mental modifications that were vital in the development of human intelligence. Rather than becoming confined to gathering and hunting at night like their quadruped competitors, hominids could walk around searching for fresh water, gathering food from wild plants, killing small animals, scavenging from the carcasses left by large predators, and holding weapons to defend themselves. With experience, selective pressure, and the evolution of intelligence, their home range grew larger as social networks and group size added important protection for the species. Their evolving intelligence equipped them to deal with a hostile environment in which they were both hunter and hunted. Standing upright, growing large bodies, developing complex brains, and using primitive tools almost 2.5 million years ago equipped hominids for extraordinary future developments and for migration to other parts of the world.[13]

Walking upright separated all species of hominids from other primates; intelligence separated human ancestors from all other hominids. Reasonable estimates exist about the growth in body size and the increase in brain size of the genus *Homo*. *Homo* had a larger brain than any of the earlier species of hominids that preceded it and its brain continued to grow rapidly. This rapid growth may have been accelerated because meat had become a significant part of the *Homo* diet.[14]

H. habilis was so named because archeologists believed that this "handy" species provided us with the first evidence of early human technology. Found in the Ethiopian Rift Valley, this 2.5-million-year-old technology consisted of hammer stones, anvils, and sharp-edged stones for cutting meat and extracting marrow.[15] Members of the species lived in the fragmented forests and savannas, and from the fossil evidence, appear to have traveled over broadly defined areas in search of food and water. Similar to other primates, they needed large amounts of drinking water to survive in the increasingly arid climate and it may have been during this evolutionary cycle that *H. habilis* developed sweat glands to cool its body. The evaporation of sweat would have required additional water to sustain body temperatures and health. With the cooling and drying of the climate, shrinking forests provided *H. habilis* with a food source of large fruits, nuts, berries, plants, and roots. As omnivores, their source of animal protein came from killing land tortoises and by scavenging the animal parts left by four-legged predators.[16]

Surviving in the midst of large carnivores and alongside other members of the genus *Homo*, including its *Australopithecine africanus* brethren who would someday succumb to the rigors of natural selection, *H. habilis*, as a member of a small but growing population, had to assess the environment daily and meet its many challenges.[17] A single miscalculation meant disaster. "Natural selection was a hard and heartless taskmaster. If hominid groups made the wrong assessment, they paid for it by death from thirst, weakness, death from starvation, increased predation, and injury or death from trauma. Since hominids adapted to the environment primarily by their wits, there was a tremendous pressure for increase in intelligence."[18]

We see this reflected in the increasing brain size in *H. habilis* and its descendants. Cranial size measurements of *H. habilis's* skull suggest a growing brain and the development of cognition, including language acquisition and the making of tools.[19] Some researchers regard the evolution of *H. habilis* as a second transcendence in the human evolutionary cycle, with the mastery of bipedalism as the first transcendence.[20] For about a million years, *H. habilis* existed in a climate that remained relatively stable. Cooler and drier remained the pattern with rather normal fluctuations. At times, the climate heated up and precipitation increased but the overall pattern was one of global cooling with less rainfall on the African savanna.[21]

About 1.5 million years ago, global climate conditions changed more rapidly and African environments grew more extreme. Vegetation, available water resources, forest cover, and sources of food familiar to hominids disappeared, changed location on the shrinking forests, or were replaced by other edible but unfamiliar plants. Changing ecological circumstances forced *H. habilis* to adapt, evolve, or suffer extinction. Changing flora and fauna would literally push *Homo* out of Africa. Evolution by natural selection would be at work again in these newer, more competitive, and harsher environments.

A diversity of species existed during this time period, with some researchers arguing that *H. habilis* coexisted with a newer species, *Homo erectus*, also identified as *homo egaster*, for almost half a million years. Formerly, the view held by most researchers was that *habilis* and *erectus* evolved in succession, one after the other.[22] Fossil remains of this newer species, *H. erectus* (named because s/he shared the erect posture with anatomically modern humans), have been potassium-argon dated at 1.8 million years in Africa, in Dmanisi, Georgia in Ukraine by 1.7 million years ago and possibly in Java, Indonesia earlier, and in Zhoukoudian, China, 500,000 years ago.[23]

According to geneticists, the evidence for these migrations is found in our DNA, those three billion nucleotides that we call the human genome. All of us share the same 99.9 percent of this genome. The remaining 0.1 percent is different and provides the evidence for migrations out of Africa to locations around the world. Tracking the identities of today's populations back to their human ancestors, geneticists noted that Native South American populations trace their lineage back to Siberia and other Asian populations. China's principal ethnic group, the Han, possess distinct northern and southern populations.[24]

What caused the impetus for this dispersal of *Homo erectus* out of Africa when its ancestors had remained there for approximately three million years without moving great distances?[25] The answer exists in the Pleistocene, beginning about 1.8 million years ago and ending with the beginning of the Holocene, the modern period, 11,600 years ago. Extreme climatic variations taxed the survival capabilities of *H. habilis* and may have caused its extinction. At times during this epoch, the African savanna's dry and cool periods became drier and colder and during wet and warm times the climate became even wetter and hotter.[26]

41

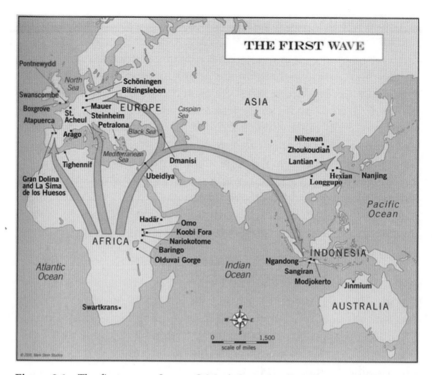

Figure 2.1 The first wave. *Source*: © Mark Stein Studios. From *Extinct Humans* by Ian Tattersall and Jeffrey Schwartz. Reproduced by permission of Nèvraumont Publishing Company.

The duration of these hot and dry extremes drove the newer species, *H. erectus*, to extend its home range in search of water and food. Increased body size, approaching anatomically modern proportions of between 100 and 150 lbs, suggests that the need for protein and bulk would force surviving members of the species to travel greater distances in search of water holes that had not evaporated in the scorching sun.[27]

Despite the hardships imposed by a harsh climate, *H. erectus* thrived by becoming the first toolmakers in a modern sense, suggesting an expanding yet still limited intelligence. They invented the Acheulean hand ax with two cutting edges to replace the more simple Oldowan tools used for a million years without much change by *H. habilis*. Unable to compete with a stronger and more intelligent species, *H. habilis* succumbed to the selective pressure exerted by *H. erectus*.

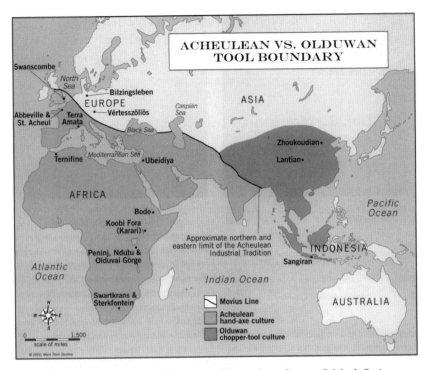

Figure 2.2   Acheulean vs. Olduwan tool boundary. *Source*: © Mark Stein Studios. From *Extinct Humans* by Ian Tattersall and Jeffrey Schwartz. Reproduced by permission of Nèvraumont Publishing Company.

Most importantly, *H. erectus* harnessed the power of fire and used it to provide heat and light. Both developments allowed them to gain more control of their environment. As omnivores, they could now cook animal and plant foods. Animal protein, whether gotten by scavenging or by the hunt, became more edible and added nutrients to the diet that improved health and longevity. Fire could also be used to ward off predators and heat their shelters. Firelight must have made *H. erectus* less dependent upon the light provided by the Sun.[28]

In an ecological sense, fires on the savanna were commonplace, especially as the climate became hotter and drier. The role of a dry and hot climate is important to contemplate when thinking about the ways in which *H. erectus* associated the elements of heat, smoke, and light. They must have watched small animals attempting to escape a fire's heat and smoke. The unlucky specimens would have become tasty morsels for these scavenger hunters. Then, cooked meat would have replaced the raw remains of large and small mammals and reptiles left by large predators.

Translating Human Intelligence into Action

The transition to a more associative kind of behavior must have been transmitted across hundreds if not thousands of generations. A larger brain suggests a capacity to comprehend and to communicate using simple commands and statements rather than gestures. Language acquisition and use probably advanced the development of cognitive and social networks.

The development of human intelligence produced the ability to receive information from one's surroundings, learn from it, and to store it in the cerebral cortex for future use. Recalling stored information, processing it in relationship to the new, and responding to the problems presented by the outside world required further cognitive development. Each new problem presented its own unique challenges to the evolving genus *Homo*. Over hundreds of thousands of years, the process of evolution by intense natural selection proceeded unabated as those members of the species who acted before thinking in this extremely hostile environment were quickly eliminated. By studying its surroundings, anticipating events before they occurred, and communicating by gesture and by voice with other members of the species, *H. erectus* probably developed a more advanced intelligence.

This stage of development represented the beginning of a common early human culture. Integrating information, developing ideas, and remembering important related events from past experiences gave *H. erectus* distinctive advantages over other species of hominids. Those who integrated quickly, learned themselves, and taught younger members the need for rapid response survived the threats posed by the world around them. They reproduced, passing this knowledge on to their offspring who understood the nature of their changing environment. They met new challenges, solved new problems, and continued the evolutionary process by natural selection. Natural selection worked on individuals within entire populations and these individuals became the evolved survivors. Living in an environment undergoing significant changes, with fragmenting forests, spreading savannas, and a cooling but volatile climate marked the evolutionary separation of *H. erectus* from other hominid species and their migration to other parts of the world.

Population Migration and Expansion

During the late Miocene (ca. 5–6 mya) the Tethys Sea was occasionally smaller than its replacement, the modern Mediterranean. As pointed out in Chapter 1, this sea dried up completely, allowing migration out of Africa into southern Europe and western Asia to take place, without crossing a major body of water. In addition, tectonic activity in geological time pushed the African plate against Europe and west Asia, creating the Alps and the mountain ranges that rim modern Turkey and Iran. Compression reduced the distances among these three significant landmasses. As a result, migration out of Africa followed an eastern rather than a western path for geological as well as climatic reasons.[29]

During the glacial Pleistocene (1.8 mya–11,600 BP), a land bridge across the southern end of the Red Sea joined east Africa to Saudi Arabia, allowing migration into Southwest Asia. Stone tools excavated at 'Ubeidiya in Israel, dated at about 1.5 million years ago, were similar to those used in the Ethiopian Rift Valley thousands of years earlier and provided evidence of this pathway into Southwest Asia. Travelers experienced a more favorable climate during the late Pliocene/Pleistocene transition about 1.8 to 2 million years ago. During the early Pleistocene, northern hemispheric glaciers would have inhibited a western path migration. Traveling eastward into Central Asia and beyond, they faced few geological or glacial obstructions. By attempting to move north, however, they would have been faced by the insurmountable Taurus and Zagros Mountains of Turkey and Iran, caused by tectonic uplift millions of years before.

Migration to new continents began in earnest with the movement of *H. erectus* out of Africa. By 1.1 million years ago, numerous locations throughout Asia contained settlement sites. The eventual expansion of this first wave reached some areas of New Oceania about 60,000 years ago and may have extended into Australia as early as 55,000 years ago.[30] Migrations occurred repeatedly, with intervals of many thousands of years. Some new migrants followed the pathway of previous travelers across Arabia through Central Asia to the coastal regions of Southeast Asia, while others crossed over into Europe. "Many anthropologists think that there were no fewer than three waves of migration out of Africa: first to the Middle East, then to Asia, and eventually to Europe. Our own ancestors, *Homo sapiens*, later re-colonized those areas and also crossed into Australia, the Americas, and the remote Pacific Islands."[31]

About 1.2 million years ago *H. erectus* may have traveled to southern Europe. Given the proximity of Africa to Europe, it seems probable that highly mobile, bipedal early humans witnessing lower sea levels would have ventured into southern Europe, while northern Europe was covered in ice. Archeologists now believe that new fossil finds represent the early incursion of humans into Europe but did not represent sustainable settlements because of its harsh glacial climate. Movement into and out of Europe coincided with the expansion and retreat of the ice. Gran Dolina, a cave in the Atapuerca hills of Spain, tells a different story, however. There, sustained occupation of the caves took place over hundreds of thousands of years and during some of the most extreme climate conditions over a 900,000-year period.[32]

Debates about the routes taken for entry into Europe remain largely unresolved. Those espousing the short route across the Mediterranean to the mainland believe that the rising and falling sea levels during the glacial and interglacial periods of the Pleistocene shortened the Straits of Gibraltar from its width of 21 miles to nine or ten miles.[33] With seawater locked in glacial ice, the separation between Tunisia in North Africa and Sicily is dotted with islands, making for a shorter journey across the Mediterranean. Failure to prove that African wildlife migrated to Europe using the same pathways, however, undermined the credibility of this shortened route hypothesis. A longer land route seems more plausible to many researchers.

Most researchers believe that *H. erectus* was stumped by the mountain ranges of Southwest Asia and did not begin the migration into Europe until discovering the northern route through Central Asia. If one accepted the possibility that humans entered Europe from the different directions and at different times, then the controversy about European migration disappears. No matter how *H. erectus* entered Europe, once they arrived there, they encountered other species of hominids competing with them for habitat and resources.

Homo Neanderthalensis vs. Homo Sapiens

After examining more than 30 skeletons located at the Sima de los Huesos site in northern Spain, researchers concluded that the earlier species *H. erectus* and the later species *H. heidelbergensis* and *H. neanderthalensis* lived simultaneously in similar European habitats. While the sharing of similar habitats was unfolding among these and possibly other species of hominids, the eventual domination by the larger brained and physically stronger *Neanderthals* overwhelmed others less suited to the severe cold climate of glacial Pleistocene Europe. The success of the dominant *Neanderthals* in Europe, beginning about 250,000 years ago, would also prove to be transitional.

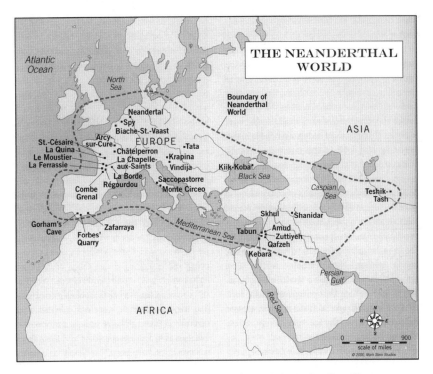

Figure 2.3   The *Neanderthal* world. *Source*: © Mark Stein Studios. From *Extinct Humans* by Ian Tattersall and Jeffrey Schwartz. Reproduced by permission of Nèvraumont Publishing Company.

One leading theory suggests that a small group of *Homo sapiens* left Africa about 40,000 to 50,000 BP, populating Asia and then Europe. Recent research suggests that *Homo sapiens* had evolved anatomically and behaviorally at least 70,000 BP while living in Africa but experienced a creative explosion once they encountered Europe's glacial climate and competition from *Neanderthals*.[34] As a separate hominid species, *Neanderthals'* replacement by *Homo sapiens* happened after many thousands of years of coexistence. Some evidence suggests that they may have even inhabited the same caves at different times in Southwest Asia.

Clearly, an expansive home range provided both with an ample supply of food, including all of the necessary animal protein, fruits, and nuts and with no competition from other species. Technologically, both species invented new stone tools and created bone carvings, as well as beads and ornaments for personal adornment, yet neither species seems to have learned much from the other.

Prior to the entry of *Homo sapiens* into Europe from western Asia 40,000 years ago, *Neanderthals* adapted to changes in the climate without experiencing excessive losses. The arrival of a new species, however, increased competition for the best hunting grounds and for the best habitats, as Europe's climate worsened. Both species responded to the competition by inventing new tools, weapons, and ornaments. The single large blade technology gave way to new modes of producing blades with handles of various sizes and function. More sophisticated stone tools with grooved teeth appeared as well as new decorative pendants and ornaments.

The combination of increased competition from a new species and a worsening glacial climate overwhelmed the *Neanderthals*. They began their retreat under this twin onslaught more than 30,000 years ago. The newcomers continued their technological, cultural, and geographical advance. *Homo sapiens* produced barbed projectile points and bone needles, painted vivid scenes on the walls of caves, and carved replicas of animals out of ivory. They buried their dead, hunted small game and large mammals, and consumed edible plants, nuts, and berries.[35]

Members of this new species excelled in ways that minimized the achievements of the *Neanderthals*. As the climate of Pleistocene Europe deteriorated further about 28,000 years ago, they retreated from their familiar homelands into southern Italy, the Balkans, the Caucasus, and to the farthest reaches of the Iberian Peninsula, to Portugal. *Homo sapiens*

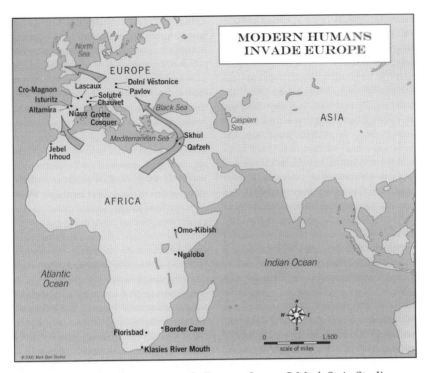

Figure 2.4 Modern humans invade Europe. *Source*: © Mark Stein Studios. From *Extinct Humans* by Ian Tattersall and Jeffrey Schwartz. Reproduced by permission of Nèvraumont Publishing Company.

extended their home range, displacing *Neanderthals* permanently. They never returned; their numbers dwindled with the advancing frigid climate and they eventually disappeared.[36]

The coincidence of their extinction with the last glacial maximum reflects the inability of *Neanderthals* to develop the cohesive social structures that sustained them earlier during long periods of climatic stress. In the past, declining material resources and dwindling food supplies caused them to retreat until the arrival of a warmer climate. Repeatedly, they withdrew and returned during their more than 250,000-year existence. They had lived through a variety of climates, including the mainly temperate cool and dry phase of 250,000–180,000 BP, the full glacial, 180,000–130,000 BP, the last interglacial with warm climatic conditions, 130,000–115,000 BP, the temperate cool period, 115,000–75,000 BP, and the cool glacial with ice sheets spreading, 75,000–30,000 BP.

However, during this last phase, *Homo sapiens* entered their domain and displaced them. The evidence regarding this displacement points to the significant changes in behavior that took place from 40,000 to 20,000 BP, a period of huge climatic stress in many parts of Europe. As population density declined, widely scattered but cohesive sites populated by modern humans emerged.

The advent of a rich and varied material culture and a developing social network distinguished *Homo sapiens* from its competitors. They imported stone and flint from increasingly distant locations for making a variety of blades. As climate conditions reached the glacial maximum about 18,000 BP, the clustering of modern humans intensified and blade technology adapted to the increasing difficulties of hunting animals in a deteriorating environment. For example, high quality flint, mined in the Holy Cross Mountains of Poland for making blades, was found in sites over 400 km (250 miles) away.

The discovery of the Venus figurines, dating from 23,000 to 21,000 BP and across a very wide geographical area, signaled the emergence of a common art form among modern humans. Advances in clothing and housing ameliorated living in austere, hostile, and unpredictable environments. Developing social relationships among peers, relatives, and trading partners minimized the effects of the glacial maximum.[37] "It is no coincidence that the figurines appeared in central and eastern Europe at a time when climates were deteriorating and food resources dwindling, and as the ice sheets expanded towards their maximum extent. For without the social alliances which such objects symbolized, these areas would have been unoccupied during the period 23,000–21,000 years ago."[38]

Developing complex social alliances across wider geographical areas, improving housing, innovating with blade technology, and storing food to avoid starvation fostered the remarkable colonization of modern humans throughout the globe. Their global expansion about 60,000 to 40,000 BP differs from the earlier dispersal from sub-Saharan east Africa of *H. erectus* a million or more years earlier. During this first expansion many geographical and climatic barriers left significant parts of the world uninhabited and resulted in distant sparsely settled communities.

Importantly, these obstacles did not deter *Homo sapiens*. With declining sea levels, they negotiated the Sunda trench from Southeast Asia to Australia, possibly as early as 55,000 BP, and reached southeastern Tasmania by 35,000 BP. Long-distance ocean travel took them to the western rim of the Pacific Ocean. They traveled over the Bering land bridge to North America as recently as 15,000 BP. Expansion through larger interconnected communities increased their prospects of survival. To protect against extinction, these new communities required a mating network of between 175 and 500 people. Stagnant for millennia, population size increased through these processes and colonization.

Early Diets and Their Nutritional Value

In the past two million years, our hunter-gatherer ancestors transformed themselves from their forebears into anatomically modern humans. Diet and climate and the relation of one to the other played decisive roles in this transformation. For most of human history, hunters and gatherers lived and ate in ways unknown to modern humans. Early humans adapted their foraging behavior in light of long- and short-term climate changes. Without knowing the effects of a diet rich in protein and minerals on brain development, they consumed high-energy food. This behavior offers the most plausible explanation for the dramatic change in the brain size of *H. erectus*, from 500 to 1,000 cubic centimeters, from the genus *Australopithecus* to genus *Homo*. Brain size grew to 1,500 cubic centimeters only as recently as 50,000 BP.

After millions of years of relatively few physiological and neurological changes among early hunters-gatherers, their eating behavior changed. According to some researchers, meat consumption by *H. erectus* was central to their physical and cognitive development. They became more efficient hunters and scavengers and carried that knowledge with them as they migrated from Africa. Other researchers argue persuasively that the nutritional value in meat consumption alone couldn't explain the dramatic physiological and neurological growth in *H. erectus*. They hypothesize that the rapid development of the species occurred with their mastery of fire and their invention of numerous stone tools and weapons. They became extensions of humanity's increasing agility and physical strength.

The question that remains at the center of the debate about early human nutrition is one of timing. When did *H. erectus* discover fire and use it to cook food? At one extreme are those who propose that the predecessors of *H. erectus*, the *Australopithecines*, began cooking tubers and roots in East Africa about 1.9 million years ago and that within several hundred generations evolved into *H. erectus*. Cooking tubers, corms, rhizomes, and other roots over these several hundred generations played a pivotal role in human evolution.

In addition to the dramatic increase in brain size from *Australopithecine* to *H. erectus*, other significant physiological changes, ranging from smaller teeth to an upright posture, took place. Some researchers argued that once again climate change in East Africa, from wetter to dryer and from hotter to cooler, reduced the availability of indigenous fruits. As a result, early hominids began collecting, cooking, and eating wild tubers and roots. "Cooking makes tubers more edible and increases their nutritional value by softening them and in many cases ridding them of toxins."[39] In addition, cooking had far-reaching social and economic effects. For example, women past the age of childbearing participated more fully in feeding their families, as they became the primary participants in collecting and cooking food.

The timing of these events also remains controversial. Some researchers believed that capturing fire for cooking occurred almost two million years ago, while others have argued that the oldest verifiable date for human fires occurred about 500,000 BP in Zhoukoudian, China, with a few sites in Europe dating from about 400,000 BP. Chinese paleoanthropologists have disputed recently, however, the claim of human fires in Zhoukoudian, arguing that the residue discovered at the site washed in from outside and may have had natural causes.[40] The timing of fire use may never be settled because the kinds of fires used for cooking were small and their residue in ash and soot vanished quickly. The direct evidence for fire use is too recent to provide clues about its use many hundreds of generations ago. "If you want hard evidence for fire in the form of stone hearths and clay ovens you are in the last 250,000 years."[41]

Cooking with fire breaks down the muscle in meat and fiber in plants to increase their nutritional value, promote digestion, and contribute significantly to physiological and neurological development in human evolution. Modern humans lived successfully for this extended period as hunter-gatherers on a highly varied diet of wild vegetation, insects, and snaring or catching small game and fish. Hunting large mammals came after the establishment of social networks and skillful methods of efficient hunting. Despite nature's bounty, the world population of modern humans remained relatively stable for this period at probably no more than a few million individuals.

As they became even more efficient hunters than their ancestors, modern humans accelerated the coming megafaunal collapse during the past 100,000 years. The climatic warming at the end of the Pleistocene about 18,000 years ago triggered the extinction. The effects of the warming ranged from coastal changes caused by rising sea levels to increased soil erosion and the collapse of important bio-systems. However, the "mega-faunal overkill hypothesis," proposed a generation ago, remains the most plausible explanation for the extinctions.[42]

Across the Earth, the extinction of megafauna coincided with human migration and settlement. In northern Eurasia, the extinctions occurred 14,400 BP and in North America extinctions took place between 13,200 and 12,900 BP. "Clinching the argument is the recent discovery of a frozen mammal on Wrangell Island in the Arctic Ocean 125 miles north of Siberia. The mammoth died only 4500 BP, which is just when the first humans reached Wrangell, and about 10,000 BP after the last Siberian mammoth perished. Elsewhere in the world, the mega-fauna disappeared as soon as humans appeared on the scene: 50,000 BP in Australia, 13,000 BP in South America, 6000 BP in the Caribbean islands, 1200 CE or thereabouts in Madagascar, New Zealand, and other Pacific islands, including Hawaii."[43]

Scientists determined what humans ate many thousands of years ago by chemically analyzing the fossilized remains of bones and by looking for trace elements of calcium. In this way, they made the comparison with modern humans in terms of their size, strength, and height. Fossilized human waste also provided evidence of human diets. Based on their exhumed remains, ancient humans were slightly taller than modern humans. Worldwide, men are 5' 5" tall and women are 5' 1" in height, on average. Our Paleolithic ancestors possessed stronger bones and lived a lifetime without dental decay. Presumably, only about 9 in 100 people lived into their 60s; those who survived the stresses of hunting large dangerous animals, infections from injuries, and the trials of childbirth had the physical attributes and the cardiovascular conditioning of modern athletes. Those who survived traumas were relatively disease-free. The modern chronic diseases – heart disease, diabetes, high-blood pressure, and obesity – were nonexistent in these hunter-gatherer societies. Remember that these ancient forebears were genetically the same as modern humans. In evolutionary and genetic terms, 40,000 years is a relatively short time period.[44]

The Broad Spectrum – an Economic Revolution

One of the most surprising outcomes of the study of Earth history is that climates in the past 150,000 years have changed suddenly. Regional changes in mean annual temperature of several degrees occurred mostly in a few centuries, sometimes decades, and less frequently in a few years. Living during periods when climate variations took place in decades or less would require humans to respond appropriately or suffer the consequences. A centuries-long "ice age" would presumably expose land bridges and allow migration before a warming period with a rising sea level blocked passage. Many other responses were also possible.

The instability of the global climate, during the final stages of the Pleistocene about 30,000 BP, allowed small autonomous groups to broaden the range of their hunting and foraging activities, ushering in what archeologists refer to as the Broad Spectrum Revolution.[45] As the last ice age entered this final stage, the climate experienced high fluctuations on timescales that ranged from decades to thousands of years. With the exception of the Younger Dryas (named for an Ice Age flower) cold period (12,600–11,600 BP), a very dry climate with equally low atmospheric CO_2 concentrations extended over large areas of the world, punctuated by dramatic advances and retreats of glaciers. High-resolution evidence from Greenland ice cores allows climatologists to plot climate changes by decade for the past 80,000 BP and monthly for the past 3,000 BP. Full glacial conditions began their final retreat about 18,000 BP and ended with the Holocene geological interglacial epoch, the current period in human history. With it came relative climate stability with variations of no more than 150 years, with the notable exception of the Little Ice Age (1300–1850 CE).[46]

Functioning in this very dry and low CO_2 environment, hunter-gatherers innovated in ways that suggest their evolved brain size and increasing cognitive abilities, the product of a million years of evolutionary development, had come into full use. They invented a number of new technologies, including grindstones to sharpen weapons and tools. They invented small projectiles, including the atlatl, a throwing stick for the hunt, as well as nets and snares to capture small mammals. Grinding instruments were invented to pulverize wild plants and prepare them for cooking and eating.[47]

Hunting and trapping small ground mammals, birds, and fish became another step in improving diet and human health. Gathering and consuming more small seeds, the predecessors of our cereal grains, also enhanced health. These improvements came at a price. The cost of hunting and trapping small game required the expenditure of more energy and time. Killing or scavenging one large wooly mammoth, however dangerous that might have been, required considerably less human effort than trapping small game that provided variety to the diet. In addition, although small animals reproduce rapidly, their numbers rise and fall unpredictably, making them unreliable as a stable source of food.[48]

In the early stages of this Broad Spectrum Revolution, humans developed a more sophisticated foraging strategy. As the range for foraging expanded and the total quantity and variety of food available for consumption increased, a more nutritious and healthier diet may have resulted. As the revolution reached maturity, however, an increasing food supply led to an increasingly dense population, as more and more individuals were required in the labor of hunting, gathering, storing, and preparing food. As the range of foraging and hunting expanded in search of a diminishing supply of small game, and as new technologies were developed to capture them to satisfy the nutritional demands of an increasing dense and sedentary human population, the Broad Spectrum Revolution entered a period of crisis. Declining game and diminishing returns from increasing amounts of human labor marked the end of this stage in human history.

Since no written records exist about the gradual decline of food resources, signaling a gradual decline in nutritional quality over time, how do we know what happened? In the Mediterranean region, the Levant, India, and western and northern Europe, humans declined in physical stature. Declining stature is associated with poor nutrition. The causes identified by archeologists and paleo-anthropologists include the increased amount of work required by foragers to secure the necessary food resources. A warmer climate may be associated with increasing rates of infectious disease and episodic stress caused by the lack of food among both an increasingly sedentary hunter-gatherer population and the small game that they hunted.[49]

The end of the Broad Spectrum Revolution coincided with the end of the Pleistocene. Accompanied by declining food supplies, associated with shifting vegetation patterns caused by climate change, it ended with increasing malnutrition and declining human height among this increasingly sedentary population. Episodic stress, in the form of seasonal hunger, an increase in the number of infectious diseases and in the rates of infection, resulted in a temporary decline in population with increased mortality and morbidity.

CHAPTER THREE

FORAGING, CULTIVATING, AND FOOD PRODUCTION

Introduction

During the late glacial Pleistocene, a hostile dry climate inhibited the transition to agriculture. Atmospheric CO_2 was so low that it stressed vegetative life, with weather extremes ranging from greenhouse to ice-age conditions within 150 years. A gradually warming climate created the conditions for agriculture and settlement. Scientists hypothesize that a decline in massive ocean ice sheets and icebergs in the northern Atlantic latitudes caused a general warming trend from 14,500 to 12,900 BP.[1] As the climate became warmer and dryer, waterways and well-watered grasslands shrank, eliminating larger hunted grassland species, especially horses and antelopes and other large species. As noted earlier, humans adapted to these losses by hunting smaller species including deer, jack rabbits, gophers, rats, turtles, and birds. Archeologists and anthropologists identify this transition as the Broad Spectrum Revolution.

A warming climate and changing amounts of rainfall improved the harvests of all growing plants for a post-Pleistocene population. Farmers captured solar energy for the production of crops and transferred that energy into the population. As historian Robert B. Marks has pointed out, "In the biologically old regime, agriculture was the primary means by which humans altered their environment, transforming one kind of ecosystem (say, forest or prairie) into another (say, rye or wheat farms, rice paddies, fish ponds, or eel weirs) that more efficiently channeled food energy to people."[2]

The interruption of this long-term warming trend and the return of the ice age called the Younger Dryas climatic episode (12,600 to 11,600 BP) reduced human foraging activities and led them to invest in labor-intensive land-clearing activities. More permanent settlements requiring less movement led to a growing human population. Archeological evidence suggests that they enriched the soil with organic materials and selected the best of the very few wild plants for eating and their seeds for planting. The severe cold living conditions of the Younger Dryas may have been one of the factors leading to the domestication of wild plants and to the beginning of food production through farming.[3] This long-term shortage of food and its impact on the population represented an important chapter in human history, for it helped to write agriculture into the historical narrative.

First discovered in the pollen record of northern Europe, the expanding glaciers of the Younger Dryas caused birch and pine forests to retreat as tundra vegetation replaced them. These changing vegetation zones and the reduction in growing-season rainfall forced adaptive responses by a human population caught in the midst of this volatile climatic transition. It triggered rapid settlement patterns in which small human communities combined to form a few large sites. With the end of the Younger Dryas and the melting of the ice, nutrient-rich soils were once again uncovered and a large amount of accessible river, lake, and stream water accelerated the transition to agriculture.[4] Cultivation and farming along with the early forms of small-scale manufacturing dominated economic development. A world of growing populations and early cities represented only a few of the remarkable changes that accompanied this change in climate.

Early Farming and a Warming Climate

Increasing numbers of farmers produced more food and the harvest often allowed more people to live in closer proximity to one another. Denser populations required organization and public order. New organizations of government based on kinship relationships and fealty and oftentimes punitive in nature controlled people and products. Disarming hostile indigenous populations and establishing armies to protect the frontiers from invasion became the single most effective early way to promote order and peace.

Harvested crops were often held in storage, protected from theft by these newly constituted armies. Kin relationships eventually gave way to more impersonal organizational decision-making. "Some of these new organizational forms were hierarchical, with more permanent, formal leadership roles instituted above the remaining population. Such leaders or decision makers, in turn, may have fostered greater concentrations of resources and labor, leading to intensified production and even larger communities."[5]

Geological changes also accelerated this transition. The geological conditions in Southwest Asia and the Levant help to explain the effect of a renewed warming climate during the Holocene and its significance in the early history of farming. The Levant is an isthmus that contains sea and desert barriers on either side. Tectonic uplift created parallel mountain chains millions of years ago. The chain of Eurasian mountains that began in the Alps across southern Europe moved southeastward toward the Taurus–Zagros arc (mountains in Turkey and Iran) and met the Himalayas in the Pamir Knot. Rising sea levels caused by glacial melt during the warmer and dryer Holocene and expanding desert margins trapped hunter-gatherers and prevented them from dispersing. In addition, the major East African Rift Valley that extends from East Africa into modern-day Jordan slowed migration. During the Holocene, the region became an abundantly watered environment. Early farmers brought wild cereals from the hills to the lowlands for cultivation. Many of the land's depressions caused by tectonic activity became oases filled with groundwater.

High evaporation rates created moisture-rich air from these emerging well-watered environments that extended from North Africa into Central Asia. The extended chain of mountains also served as a convenient barrier to the westerly winds. As the westerly winds crossed the desert region in the eastern Mediterranean they dropped their moisture on the mountains, creating the Fertile Crescent, a macro-oasis with run-off from the mountains feeding the river systems of Mesopotamia.[6]

Mediterranean vegetation began to expand into Southwest Asia, particularly the Levant, beginning about 12,500 BP. Cereals, pulses, oak, almond, and pistachio trees that foragers had picked in the lowlands as early as 20,000 BP now began to appear in the hill zones of the Fertile Crescent. The effect of an average temperature change of 9 °F provided an optimal environment for plants formerly restricted to the lowlands. The warmer elevated terrain and the virgin basaltic soils created by tectonic uplift increased yields. The growing season at higher elevations expanded by as much as five to six months. Wild plants retreated as migrating farmers engaging in cultivation moved to higher and higher elevations in response to a warming climate.

With the return of climate warming, glaciers receded, air circulation patterns shifted, and vegetation zones expanded. Emptied for hundreds and sometimes for thousands of years by mile-high glaciers, deeply cut river valleys began to fill. Human groups settled in rich zones of open forests and in areas where they could domesticate wild plants. Geological rifting, climatic disruptions, and environmental changes led to rapid human economic and cultural adaptations.

Settlement and Domestication

Rising temperatures created an extended growing season. The presence of abundant wild plants attracted foragers and encouraged settlement. Once foragers accumulated sufficient food, they settled down and limited their movements. The longer the growing season, the longer they remained in place. Weeding vegetation, selecting the most nutritional-looking plants for eating, inadvertently added biomass to the soil as garbage and excrement changed the genetic composition of the soil and plants. Domestication through natural selection may have become an accidental consequence of human activity during this early period of cultivation.

In *After The Ice: A Global Human History*, Steven Mithen described the genetic differences between wild and domestic plants. For wild cereals growing on the forest steppes of Southwest Asia during this transition to a warmer and wetter climate, an important genetic change had to occur for them to become domesticated. When mature, brittle ears of wild cereals burst and fall to the ground. Once domesticated, the ears of grain cling to their stocks waiting to be harvested, so they cannot reproduce themselves without human intervention. Humans intervene precisely at the moment of maturity by beating the grain and dislodging the ears into their waiting baskets.

The most important result of human intervention is the mutation of a single plant gene that turns wild cereals into domestic varieties. This same process occurs in various strains of peas and lentils.[7] As we will note throughout this chapter, humans have engaged in the genetic modification of foods since the transition from hunting and gathering to agriculture millennia ago. Only recently has the process become consciously scientific with advances in plant genetics.

Settlement also promoted further developments in the division of labor, with some workers using their knowledge of grinding stone technology to produce mortars, pestles, seed grinders, and ground-stone cutting implements. Although large numbers of seasonal settlements existed for thousands of years during the Pleistocene in Europe and Eurasia, a clear relationship existed between climate change, settlement, and the extensive distribution of wild plants, particularly in Southwest Asia, during this early phase of agricultural expansion.

The process of producing food about 12,000 BP did not appear suddenly with hunter-gatherers "discovering" the benefits to farming as opposed to foraging. From a labor and risk perspective and without the promise of a nutritional return from a new method of procuring food, farming became an unpredictable and difficult transition. According to a co-evolutionary process, the domestication of wild plants, farming technology, and population density led naturally to an increased yield of these plant foods. "This increased yield stemmed from evolutionary changes, through biological selective pressures, as a result of expansions of disturbed ground within a region, increased numbers of human beings and increased sedentism, operating in concert."[8] Human–plant interactions served as the key to understanding the transition from forging to farming. Changes in the land, through the improvement of soil conditions by human and animal waste disposal and the clearing of unwanted vegetation around edible wild plants, transformed plants into cultigens.

Since hunting and gathering were significant subsistence activities, local inhabitants became sensitive to their microenvironments. As they increased plant food consumption, they became aware of seasonal changes and the fluctuations in the availability of food that marked these variations. Small bands of families congregated into larger tribal groupings during the wet spring and summer months to hunt smaller game and gather plant food. Some evidence exists to suggest that they consumed wild squash, chili, and avocado. All of these would later become domesticated.

As the benefits of combining into larger groups became known, semi-permanent settlement patterns of migrating groups became more stable. "More plants were utilized, including cultivated and domesticated species. Maize appears in the archaeological record during this phase and the relative quantity of meat in the diet declines."[9] By this time, the amount and number of plants increased, suggesting a stable and sufficient food source. Changing dietary patterns, in turn, were related to demographic growth. While mobility reduced the number of children that hunting and gathering communities can easily transport and nurture, sedentary behavior placed fewer such restrictions on the group.

Hunter-gatherer women engaged in strenuous exercise. Regularly moving from one place to another made it impractical to cook foods digestible for infants. So, prolonged breastfeeding became a common practice. Both activities temporarily reduced fertility.[10] "Earlier weaning and shorter birth intervals [were] associated with the availability of soft, palatable, digestible weaning foods ... the weaning foods were probably the highly nutritious, indigenous starchy-seeded annuals. Thinner, more efficient ceramic vessels facilitated their preparation through boiling."[11] Thus, tool technology and ceramics transformed settlement patterns, and as complex and permanent settlement patterns appeared, population size rose.

Some wild plants had probably evolved genetically into cultigens through the evolution of naturally selected traits favored by foragers as they cleared vegetation to more easily identify these desirable and edible plants. Possibly all of the above and much more unfolded long before humans began to plant seeds and consciously prepare the land for farming. They continued to forage for edible plants, selecting the best, discarding the seeds in dump heaps, and destroying the worst. Hunter-gatherers may have been attracted to some wild plants and not others. For example, plants that produced edible vegetables or potential containers such as gourds and spices fit this category of attractive plants.

In some locations their gathering practices led to widespread devastation. For example, in western Scotland about 8700 BP, the harvesting and roasting of approximately 100,000 hazelnuts led to the collapse of the hazelnut woodlands in the region. According to Steven Mithen, "these hunter-gatherers were certainly not living 'in balance' with nature."[12] In addition to this kind of pseudo-agriculture, foragers who lived in proximity to farmers learned farming practices from them.

Early Agricultural Communities

When did the planting of seeds for food and the harvesting of crops for subsistence become intentional and when did settlement patterns of foragers change to facilitate cultivation? Around 9000 BP, wheat and barley grew in Southwest Asia from Anatolia to Iran. Peas, lentils, chickpeas, broad beans, and flax (from which the first linen cloth was woven) followed later in the same region. The archeological site of the city of Catalhoyuk, Turkey, originally excavated in the 1980s, seems to prove the theory that settlement and agriculture took place simultaneously. At its peak about 9000 BP, the city contained as many as 10,000 people and was a third of a mile across in size. Because the site contained cattle bones, as well as murals and sculptures depicting animals, archeologists concluded that farming and the domestication of animals, two related activities, proved the theory that agriculture, domestication, and settlements were synergistic activities.

Today, the site of Catalhoyuk has become the focus of a more detailed archeological investigation. Preliminary findings offer a counterpoint to the familiar historical narrative about the close relationship of agriculture, the domestication of animals, and settlement patterns. Many of the animal bones excavated were those of wild sheep and goats. The cattle may have been either wild or domesticated since their bones reveal little about the usual reduction in size that takes place with domestication. Since smaller animals are easier to control, herders bred size and hostility out and selected smaller and more docile cattle for domestication. Evolution by natural selection as well as a more restrictive diet and changed patterns of mobility resulted in smaller cattle, a sign of the transition from wild to domestication.

Archeologists used another method to determine the extent to which agricultural practices replaced hunter-gathering activities. While herders killed mostly young males and kept a large breeding stock of females, the method involves studying the ratio of old to young animals and males to females. By studying large numbers of bones, they determined the ratio of old to young and males to females at an archeological site. At Abu Hureyra, Turkey, 11,500 to 10,000 BP, the early inhabitants lived on wild seeds and hunting. DNA analysis of microfossils indicated that wheat was grown there about 11,000 BP. The microscopic examination of human teeth, however, showed no sign that the early inhabitants were eating large quantities of grain.[13]

At another ancient site, in the Yangtze (Chang Jiang) River Valley of China, microfossils from 13,000 BP indicate that hunter-gatherers cultivated rye and rice, 4,000 years before rice became a staple in the Chinese diet. Since permanent settlements did not appear there until about 8500 BP, a gap of 4,500 years separates cultivation from human settlement. In Coxatlan, Mexico, 9500 to 7400 BP, hunter-gatherers divided their time between hunting camps and larger villages where they gathered wild plants. The evidence mounts about gathering, breeding, and cultivation around the world. In Israel, about 17,000 BP wild grains were a major food source; rice was cultivated in Syria 15,000 BP and by 14,000 BP squash was cultivated in Ecuador and, 2,000 years later, in Mexico. Almost 6000 BP maize cultivation began in Panama.

In the New Guinea highlands, for example, archeologists discovered evidence of the deliberate planting of taro and bananas sometime between 6950 and 6440 BP. These were plants indigenous to the island without evidence of influence from Southeast Asia. Agriculture emerged independently without becoming linked to rapid population growth, incipient urban growth, political institutions, and social hierarchies. Despite the advent of agriculture in New Guinea, life there remained non-hierarchical and relatively egalitarian. "The evidence of early agriculture in Highland New Guinea signifies the potential diversity of prehistoric trajectories following the inception of agriculture, and challenges uni-linear interpretations of human prehistory."[14] The invention of agriculture was a complex set of events. In some places, cultivation took place thousands of years before permanent settlements became integrated social and economic organizations.

Throughout the world, the prominence of river valleys during the Holocene transition cannot be underestimated. These newly emerging floodplain ecosystems, rich in fish and waterfowl protein, provided a reliable source of food for hunter-gatherers. They attracted human hunters long before plant cultivation became an important human activity. The ebb and flow of the watersheds created main and tributary channels, swamps and meadows, and naturally drained levee soils that were annually replenished by the silt from upstream winter ice and snow melt. These nutrient-rich soils would become valuable natural resources for Neolithic (New Stone Age) farmers almost everywhere. Despite its climatic fluctuations, the Holocene transition created the conditions for the transition to agriculture and a protein-rich diet.

Although explanations about the causes for the transition from hunting and gathering to agriculture abound, within a few thousand years agriculture began in several locations. In regions with a distinctive dry season, farming was relatively easier than in regions with heavy precipitation and cloudy weather. Although they were all alluvial floodplains, periodic droughts in these floodplains created stress on plants and animals that required consistent feeding. The earliest domesticated annual plants flourished during periods of seasonal drought. These plants included annual grasses, legumes, or cucurbits, which manufacture unusually large seeds, a characteristic during periods of seasonal drought. The existence of large-seeded annuals may explain the transition to agriculture because farmers cultivated them more easily than small-seeded wild plants.[15] Storing food for humans and animals and securing seeds became priorities for early farmers faced with these unpredictable climatic conditions. The highly variable climate of the early Holocene may have provided the environmental conditions for the cultivation of the wild precursors of domesticated wheat and barley in the world's earliest agricultural centers.

Early Agriculture in China

In regions of the world where seasonal droughts occurred during a cold winter rather than a warm winter, other environmental factors intervened to support incipient agriculture. Weak seasonal droughts in northern China explain, for example, "the huge, highly erodable loess plateau [that] would have supported dense stands of foxtail millet naturally, and it contributes such a heavy sediment load to the Yellow (Huang He) River that wild millet would also probably have covered large areas of the substrate laid down each year in the enormous floodplain. In all probability, wild rice also was present in the early Holocene, when temperatures were warmer than today."[16] Once again, the phenomenon of floodplains exerted a significant environmental influence on the development of early agriculture and the eventual growth of settled populations.

Rivers overflowed their historical banks during seasonal flooding, leaving rich sedimentary nutrients for growing wheat and barley. Each year seasonal flooding brought additional nutrient-rich soils to the valleys and plains. "Even today, these crops are cultivated on the floodplain of the Indus without ploughing, manuring, or providing additional water."[17] For five to six thousand years, humans practiced and spread agriculture by exploiting the naturally fertile floodplains of Eurasia's major rivers, the Old Kingdom of Egypt, the Sumerian civilization of Iraq about 7300 BP, and in Mahrgarh, a small farming village located west of the Indus River Valley, between 9000 and 7500 BP.

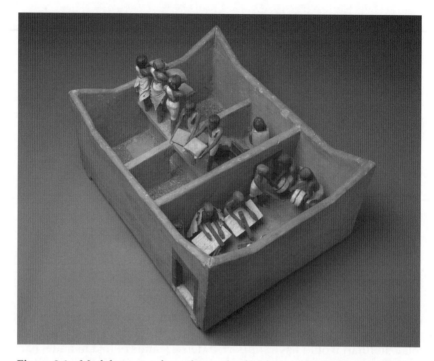

Figure 3.1   Model granary from the tomb of Meketre, Middle Dynasty 12 in
Ancient Egypt. *Source*: The Metropolitan Museum of Art, Rogers Fund and
Edward S. Harknes Gift, 1920 (20.3.110). © The Metropoliton Museum of Art.

Over the millennia, climate conditions changed significantly in each
of these civilizations, as climate systems in the regions became even
warmer and dryer. Nearly all centers of early agriculture experienced
extreme seasonal drought. As the climate changed and Holocene seasonal temperatures became more extreme, plants and animals were
forced to live closer to humans, promoting domestication.

In China, little evidence exists of agriculture dating back to 10,000 BP
but a number of cultural groups farmed northern China's central plains
from about 8500 to 7000 BP. During the late Neolithic period about 7000
BP, these northern Chinese farmers cultivated a variety of crops, including some paddy rice, hemp, soybeans, Chinese cabbage, and leaf mustard. Even so, millet remained the dominant crop throughout northern
China during this period. At one site, identified as belonging to the
Cishan culture (8000–7500 BP), archeologists excavated 345 rectangular

grain storage pits in 1984, eighty of which contained decayed foxtail millet or millet husks. If we converted these decomposed remains into fresh millet, it might have weighed as much as 110,000 pounds. By including the potential grain storage of the empty pits, the total becomes 220,000 pounds. By any measure, this storage space exceeded the capacity of all previously excavated sites.[18]

Climatic conditions strongly influenced Chinese cultivation patterns. In the temperate northern plains with four clearly distinct seasons, harvesting and storing crops for the harsh winter months became primary agricultural activities. In south China, however, a different history of cultivation took place. With no winter season, ample precipitation, and an environment rich in animals and plants, farmers experienced no pressure to domesticate wild specimens. Although inhabitants of the region observed the natural growth cycle of plants, including wild rice, and recognized the nutritional value of rice, they would have no need to cultivate it until a combination of climatic change and demographic pressure required it. Only then would agricultural activity that had reached an advanced stage in northern China penetrate the south. In China and other core Far Eastern regions at latitudes between 10 and 15 degrees north, farmers grew domesticated rice by 8000 to 7000 BP.

During the same millennium, rice farming appeared throughout most of Southeast Asia, and 2000 BP rice cultivation entered Sumatra and southern Sulawesi in Indonesia. By 4500 BP, wheat farming appeared further eastward from Southwest Asia to the western region of the Indus River. Rice and sorghum farming had also reached the Indus Valley 4000 BP and later in the millennium these crops were growing in most of Pakistan.

Early Agriculture in Africa

In sub-Saharan Africa, the transition to agriculture was as complex as it was anywhere in the world. Early farmers domesticated more than a dozen local plants, ranging from sorghum cultivated from Lake Chad to the Nile, millet from Ethiopia to northern Uganda, African rice in the middle delta of the Niger River, to tiny-seeded teff in Ethiopia and yams in West Africa. During this transition, herders imported goats and sheep from Southwest Asia. They imported domesticated cattle either from northern Africa or from Asia. As a result, domesticated livestock

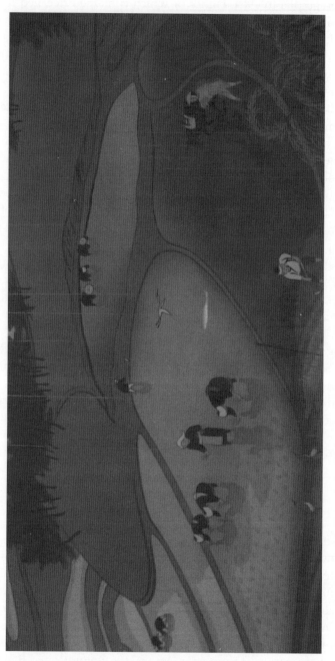

Figure 3.2 Rice cultivation in the Kanto. *Source*: Permission by the owner, Colin Sargent.

were present in sub-Saharan Africa 7000 BP.[19] By around 6500 BP, wheat had moved south from Egypt into southern Ethiopia. Millet and yams grew from eastern Sudan to northern Nigeria by 5000 BP and spread southward to Kenya and Tanzania.

Early Agriculture in Mesoamerica

The origins of agriculture in the environmentally diverse region of Mesoamerica differed from agricultural beginnings in other parts of the world for numerous reasons. First, the region lies in the tropics extending from northern Mexico to Costa Rica, with the variety of altitudes shaping the region's climate. In most areas a climatic mosaic exists with predominantly cold, temperate, and warm temperatures.[20]

Ecologically, rainfall sustained the region's agriculture. Early farmers cultivated a few selected plants from their wild ancestors. Over time, the production of maize, beans, squash, gourds, chili peppers, and avocado would shape the economies and social fabric of the region. A pattern of diffusion would be played out throughout the Americas as farmers grew beans, potatoes, and peppers from 8000 to 6500 BP. Between 8000 and 6000 BP, they cultivated squash in south-central Mexico, and by 5000 BP the growing of squash reached South America.

With the domestication of maize in the same region 6500 BP, food production took a major step forward. By 5000 BP farmers in northern Mexico grew maize. Around 4000 BP, maize appeared in the southwestern United States and in Peru, spread into by Pennsylvania by 3000 BP, and by the beginning of the Common Era reached the upper Mississippi-Ohio valleys. Both maize farmers and their ideas about cultivation spread through the hemisphere, replacing hunting and gathering societies. Root crops, including yams and tubers, eaten by hunter-gatherers for hundreds of generations, continued to grow in their regions of origin. Yams, for example, were grown in cultivated land 10,000 BP in West Africa and Southeast Asia.[21]

Early Agriculture in Europe

As early farmers migrated from Southwest Asia into the heartlands of Europe, they carried their seeds, plants, and technology with them.

They settled in the central and western Mediterranean region, and constructed houses with timber frames and bred their domesticated cattle and sheep. They grew crops, and made pottery and polished stone instruments – axes, grinding stones, knives, and weapons. Between 8000 BP and 6500 BP these Neolithic farmers had cultivated the land of its Mesolithic hunters and gatherers (10,500–6000 BP). Agriculture spread throughout Europe by the combined efforts of the newly arriving Neolithic farmers and the adoption of their know-how by the continent's Mesolithic inhabitants whose culture was overwhelmed by migrating farmers. Only in the extreme northern reaches of the continent did hunting and gathering continue into the modern era but it finally ended there as well.

Further north in central Europe, beyond the immediate domain of farmers, Mesolithic people continued to hunt wild boar and deer, and make stone tools. When the hunt failed them, however, they adopted cereal cultivation and animal herding. In addition, they were aided by the dispersal of farming technology throughout Europe by acquiring pottery-making from these early farmers for efficient cooking and storage. Simultaneously with these developments, advancing farmers spread wheat quickly across Eurasia. By 8500 BP, it grew in agricultural communities in the eastern Mediterranean from Iraq to northern Greece. From 8000 to 7000 BP, wheat moved through the rest of Greece and as far north and west as the Danube River regions. By 6500 BP, it reached northward to Poland, westward to Germany and Sweden, and 500 years later to the British Isles.

The spread of farming throughout Europe was aided by changing climate conditions. Between 6000 and 5200 BP, global warming and the spread of farming resulted in the production of cereals in central Europe for the first time, identified today as the Magdeburg–Cologne–Liege region. The spread stopped with a cold period, especially between 2900 and 2300 BP. Centuries of volatile weather conditions oscillating between the extremes of hot and cold, wet and dry, were followed by a warming trend that lasted from 1000 CE to about 1300 CE. Identified by climatologists as the "medieval warm period," agriculture spread more rapidly into central and northern Europe.

Along with the ameliorating climate, the axe and the iron plow transformed the land's ecological balance. The process was accelerated by the removal of forest cover and its replacement with land cleared for seeding. Pollen analysis in these regions showed that forests indeed retreated during this warming phase. Warming also permitted the planting of crops into the highlands and mountain districts of central Europe by as much as 60 to 70 feet in elevation. "On the hills of northern England, the plow reached its highest levels in the twelfth and thirteenth centuries – beyond the limits of the wartime emergency plowing campaign of 1940–44."[22]

With the arrival of the Little Ice Age (1300 CE–1850 CE) in the northern hemisphere, affecting Europe and its colonies in the Americas, climate change retarded agricultural expansion and productivity. Pollen evidence gathered by scientists showed that the cultivation of grains and fruits had already retreated from the higher elevations in central Europe by as much as 60 to 70 feet as early as 1300 CE. Severe cold temperatures created a subsistence crisis in Europe and in parts of the Americas as widespread hunger threatened the growing settled populations.

World Agriculture During the Age of Manufacture and Industry

From 1750 CE to 1850 CE, farmers faced volatile weather systems that disrupted planting and harvesting. Killer frosts in the fall and spring wiped out burgeoning crops. A series of volcanic eruptions that sent millions of tons of smoke and ash into the atmosphere, blocking the sun's radiation from reaching the surface of the earth, accompanied the deteriorating weather. Eruptions in Japan and Iceland in 1788 CE and 1789 CE and on Mt Tambora, Indonesia, in 1816 CE lowered worldwide temperatures, reduced global crop yields, and caused widespread hunger.

Throughout Europe, during the age of manufacturing and industry, farmers struggled to meet the dietary needs of a growing urban population and an expanding rural population. Although the world economy was primarily agricultural with almost 95 percent of the population engaged in farming, rural peasants and urban paupers faced the regularity of food shortages. The masses of rural and urban dwellers generally ate little and poorly, even when times were good. Despite their meager incomes, they spent upwards of 60 to 80 percent of their incomes on nutritionally poor grains.

According to Carlo Cipolla, "The ultimate reason for which grains are the food of the poor must be sought in the ecological chain of energy.

The wheat plant transforms solar energy into chemical energy. Animal meat, instead, is the product of a double transformation process in which are added the 'losses' of the primary process, connected with the growth of forage, and the 'losses' of the secondary process connected with the development and growth of the animal."[23] Additionally, the unpredictability of the harvest caused prices to fluctuate wildly. Pest infestation, an ever-present threat to crop yields, exacerbated by extreme weather patterns of the Little Ice Age, affected the harvest and contributed to unstable prices.

Poor harvests and soaring prices limited the purchasing power of the majority. An undernourished population under the best of circumstances faced hunger constantly. In the worst of circumstances, poor harvests led to starvation. So, despite the spread of agriculture for the thousands of years before the modern era, hunger and famine plagued populations everywhere. Primitive farming technology, poor systems of transport, and the absence of fertilizers added to the burden of producing enough food for a growing settled population.

The escape from hunger and an early death became a real world conundrum. The lack of nutrition made it difficult to work. Physical weakness caused by an inadequate food supply inhibited the kind of work needed to plant, sow, and harvest food crops that provided the energy to work more. Our ancestors were caught in a nutritional trap with no easy way out. To sustain themselves biologically, they adapted and became shorter and thinner than their hunter and gather forbearers. By getting smaller and requiring less nutrition, they could survive and work but the cost of nutritional deficits was a shortened life.

Malnutrition also compromised the immune system and the lack of food meant that weakened individuals were susceptible to infectious diseases. Again, individuals faced another challenge since disease interfered with the metabolism of food into energy. Cleaning up the environment, especially the water supply, by eliminating human and animal waste and preventing debilitating gastro-intestinal diseases, cholera, typhus and typhoid fever, promoted health and a longer life.

Chapters that follow will provide additional details about the impact of disease on population as rapid urbanization and industrialization eclipsed efforts to insure and maintain a sanitary environment and promote public health. During these periods of economic growth in nineteenth-century Europe and the United States, public health suffered and average life spans were shortened.[24]

Figure 3.3 Detail of *Les tres riches heures*. An early example of crop rotation to renew the land's fertility. *Source*: From Wikimedia Commons, the free media repository.

Despite these barriers to growth, evidence of small increases in agricultural productivity and population began about 1300 CE.[25] By 1500 CE, the threat of starvation waned somewhat with the expansion of all sectors of the European economy, with agriculture growing the most rapidly. In central Europe, grain production grew rapidly on large landed estates and was shipped by way of the Baltic Sea to many western European cities. With the growth of cities, farmers cultivated more arable land and intensified their production by expanding field rotation. Farming productivity grew along with the creation of a connected trading network. Growth was spurred on by farmers specializing in commercial crops, mixing crops to various soils, and on increasing the number of livestock per acre to fertilize the land. By adding animal protein to human diets, the overall wellbeing of the continent's growing population improved. The Little Ice Age interfered with much but not all of this growth.

The causes behind this increase in farm productivity have been the subject of much debate among historians. Most agree, however, that farmers responded to the demand for food from an increasingly urban non-farm population involved in other forms of productive work. The market-based opportunities began to appear as early as the 1500s and accelerated after the beginning of industrialization in the 1750s. By shifting their production toward crops for the growing urban market, investing in livestock and land, and working harder, farmers experienced a rise in productivity and in incomes.

In addition, the exploration of the New World in the 1500s by Spanish and Portuguese conquerors and settlers added a number of crops to the global food stock. As white and sweet potatoes, corn, beans of many varieties, cassava, squash, and many more became staples in the diets of Europeans, Africans, and Asians, the world's population began to rise significantly during the past three centuries. In addition, these New World crops grew in regions, elevations, and on soils not inhabited by the Old World staples of wheat, rice, barley, and oats. To defeat age-old problems of declining yields on fields cultivated year in and year out, farmers employed the familiar pattern of crop rotation, turning farmland into pastures to replenish the land's natural nutrients. New World crops added food to the population's diet without diminishing the yields of traditional staples.

During the centuries that followed, potatoes brought from the Andean highlands by Spanish explorers spread across Europe in the sixteenth century, reaching Imperial Russia in the early decades of the nineteenth century. In the last 40 years of that century in Russia, potato production grew over 400 percent.[26] The adoption and spread of maize in Europe was equally stunning. Because its cultivation requires a hot dry climate, corn spread across southern Europe from Portugal to Ukraine. Although mainly used as food for livestock, millions of Europeans added corn to their diets during the past 400 years.[27]

In Africa, where few of the world's cultigens originated, importing food from Asia and the Americas sustained its growing population in the past four centuries. Maize, cassava, peanuts, squashes, pumpkins, and sweet potatoes fed millions of Africans. Maize was cultivated as early as the middle of the sixteenth century and by 1900 was found almost everywhere. It even competed with the production of millet and sorghum in the drier parts of the continent.[28] Despite the New World's voracious exploitation of Africa for slaves, the growth of the African population across these past centuries can probably be explained by an expanding food supply.

Although the cultivation of New World crops provided more food for growing Asian populations, agriculture in China, India, and Japan was so productive that food costs there were much lower than in Europe. "In India (and China and Japan as well) the amount of grain harvested from a given amount of seed was in the ratio of 20 : 1 (e.g., twenty bushels of rice harvested for every one planted), whereas in England it was at best 8 : 1. Asian agriculture thus was more than twice as efficient as British (and by extension European) agriculture, and food – the major component in the cost of living – cost less in Asia."[29]

Not only were food costs lower in China but its agricultural output also exceeded that of European farmers. In southern China two to three harvests a year on the same plot of land produced enough food for the country's growing population of 140 million in 1650 and about 390 million in 1850. Innovations in irrigation, fertilizers, and insect control insured bountiful harvests. Upon visiting China in the 1720s, a French traveler asked the probing question: "Do the Chinese possess any secret arts of multiplying grain and provisions necessary for the nourishment of mankind?"[30]

The introduction of New World crops served to enhance the competitive advantage of Asian agriculture over European, lowering the cost of living and making Asian wage rates in industry higher than those in Europe. The reasons for the reversal of these conditions during European industrialization will be explored more fully in a future chapter.

The Green Revolution

This global diffusion of food occurred without major technological innovations such as mechanical harvesting or the use of commercial fertilizers. By 1850, population size in Europe and Russia had doubled to about 275 million, and farmers planted a rising share of the arable land in wheat for a growing commercial market.[31] Globally, Asia's population rose from about 327 million in 1650 to 741 million in 1850; much of the growth, it can be argued, was caused by the impact of food crops from the Americas.[32] By the twentieth century, food plants from the Americas, mainly corn and potatoes, made up about 20 percent of China's annual food production.[33]

The agricultural bounty and population expansion continued with the impact of fossil energy for machinery, fuel, fertilizer, and pesticide

Figure 3.4 Plowing an alfalfa field by tractor. *Source*: From Wikimedia Commons, the free media repository. Scan of Collier's New Encyclopedia, Volume 1 (1921), opposite p. 58, panel H.

production. In the early nineteenth century, the world's population probably reached one billion, and by 1903 it stood at 1.625 billion, rising at an annual rate of 0.5 percent. In the twentieth century, the world passed six billion people with an extraordinary annual growth rate of 1.3 percent. To accommodate this spectacular twentieth-century surge in population growth, a 3.7-fold increase, the world's food supply expanded sevenfold.[34] Unlike much of the past history of agriculture in which population size outstripped crop production, twentieth-century farmers produced more food per person than at any other time in human history. Hunger remained a global problem because of an unequal distribution of food, the rise of mono-cropping, and the disappearance of family farms devoted to producing a range of food crops for local consumption.

As environmental historian J. R. McNeill has pointed out, plant breeding became synonymous with the twentieth-century Green Revolution. With the introduction of gas-powered machinery, farming practices changed in significant ways for the first time in a thousand years. "The spread of gas stations, repair shops, and mechanics provided farmers with the needed support system."[35] The conscious genetic engineering of plants symbolized by the Green Revolution represented a necessary step in the long-term quest for agricultural intensification. Along with the replacement of organic fertilizers in the form of manure and crop residues with synthetic fertilizers described more fully below, large-scale twentieth-century monoculture farming may become unsustainable.

The Green Revolution owes its beginnings to the work of many scientists. Gregor Mendel's (1822–84) work in mathematical genetics was rediscovered decades after its publication. Charles Darwin (1809–82) published his experiments in producing hybrid corn in 1876, and a number of agronomists well into the twentieth century in Japan, Mexico, the Philippines, India, China, and elsewhere continued these experiments. They focused on staple food crops: corn, wheat, and rice. McNeill described one such experiment sponsored by the International Rice Research Institute, founded in 1960 with financial support from the Rockefeller Foundation, that had a far-reaching impact on food production in Asia. "Using dwarf rice strains first selected by Japanese breeders in the 1920s in Taiwan (then a Japanese colony), rice geneticists created high-yield rice varieties that combined the best features of tropical (*indica*) and temperate (*japonica*) rice."[36] Later, plant breeding experiments with rice spread throughout south Asia from India, Indonesia, and Korea.

The effects of plant breeding to produce high-yield staples resulted in a global cornucopia of food that, by the end of the 1990s, represented more than three-quarters of the developing world's agricultural production of wheat, rice, and corn. Once again, J. R. McNeill provided the larger context for this global phenomenon: "This dissemination of new breeds amounted to the largest and fastest set of crop transfers in world history. With its basis in crop breeding and the transfer of successful strains, the Green Revolution merits comparison with the great historical crop introductions, such as the arrival of American food crops (maize, potato, cassava) in Eurasia and Africa after 1492, the importation of Southeast Asian plantains into tropical Africa and the Arab introduction of citrus and sugarcane to the Mediterranean world after 900 CE."[37]

Increased mechanization, high-yielding hybrid crops, and a new generation of twentieth-century fertilizers led to massive global harvests. Without the industrial synthesis of ammonia from nitrogen and hydrogen, however, and the resulting production of fertilizers to replenish the soil's nitrogen nutrients, nature would have imposed production limits on twentieth-century agriculture. According to the geographer Vaclav Smil, "synthetic fertilizers have become the dominant source of nitrogen in modern, intensive cropping, supplying at least 60%, and commonly 70–80%, of the nutrient reaching intensively farmed fields."[38] How did this transformation of twentieth-century agriculture take place?

The use of chemicals for wartime purposes reached its zenith during World War I, as allies and their enemies tried to vanquish each other by using airborne chemical agents. Narratives about the use of "mustard gas" to immobilize one's enemies became commonplace as combatants became bogged down during years of trench warfare, causing the deaths of millions of soldiers. Two German chemists, Fritz Haber and Carl Bosch, discovered the formula for combining nitrogen with either oxygen or hydrogen and producing a compound (ammonia NH_3) that would regulate the growth and metabolism in plants. Although nitrogen is the most prevalent element in the atmosphere, it is unusable to plants in this form because of the stability of its internal bond. It needs to be destabilized in order for it to benefit the growth of proteins in plants. By figuring out a way to create a destabilized form of nitrogen, Haber and Bosch set the stage for the revolution in twentieth-century agriculture.[39]

As a result, intensively farmed land producing cereal grains improved their yields from 0.75 tons for every hectare in 1900 to 2.7 tons in 2000. Not only were these higher yields necessary to sustain the twentieth-century population explosion, but also nitrogen fertilizers promoted intensive agriculture on less land. Among the positive outcomes related to the use of fertilizers has been the reduction in agricultural land use, which has resulted in forest renewal, grasslands recovery, and the expansion of wetlands in Europe, North America, and much of the developing world.

The production process for nitrogen fertilizers is a costly one, however. Destabilizing a tightly bonded chemical like nitrogen requires high-energy inputs. The energy from burning more than a ton of coal is required to produce a mere five and a half pounds of nitrogen fertilizer. When the cost of fuel is cheap, few give much attention to this factor.

Figure 3.5 Industrial chicken coop. The mass production of chickens using feed enhanced with growth hormones and antibiotics to prevent the spread of infection. *Source*: From Wikimedia Commons, the free media repository.

However, in a twenty-first-century world of increasing consumption of non-renewable resources such as coal and oil, the productivity of the land is linked inexorably to the price of energy.[40]

In addition, the dependence on synthetic nitrogen fertilizers disconnected crop production from animal husbandry. The organic linkage of cycling nitrogen-rich manures and plant residues back into the land for crop growth maintained the land's ecological balance. Replacing organic wastes with synthetic compounds reduces the soil's organic matter, compromising its quality and leading to compaction, runoff, erosion, acidity, and eutrophication. In other words, the overuse of synthetic compounds disrupts the Earth's ecological balance and places increasing burdens on the land, water, and air.[41] In terms of air quality alone, the effects of nitrogen fertilizers are noteworthy. As a volatile

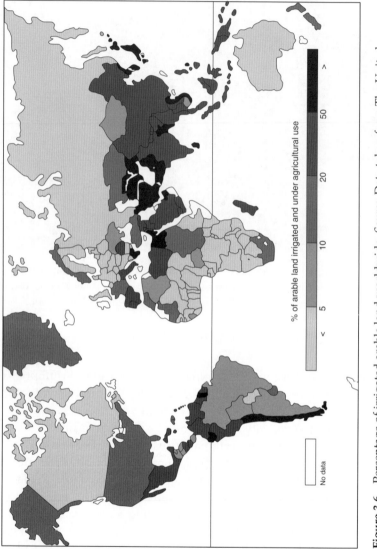

Figure 3.6 Percentage of irrigated arable land worldwide. *Source:* Data taken from The United Nations Food and Agricultural Organization, 2003. Map constructed by Colin Sargent.

compound, nitrogen applied to the soil degrades into gaseous nitrogen oxides that become heat-trapping greenhouse gases more powerful than carbon dioxide. Atmospheric ozone depletion and surface-level ozone pollution are linked to concentrated levels of nitrogen oxides. As a result, a connection can be made between agricultural uses of fertilizers and global pollution and warming.

In 1972, the ecologist John McHale, noting that one-sixth of the world's food supply was dependent on the use of synthetic nitrogen, stated that: "There is a tendency to separate agriculture from industry in everyday thinking, but the image of the farmer as conserver, and industry as spoiler, of nature is no longer true – if it ever was. To make each million ton of such nitrogenous fertilizer annually, we use, in direct and related industries, a million tons of steel and five million tons of coal. Some 50 million tons of such support nitrogen are estimated to be required annually by 2000."[42] Not only was McHale's prediction accurate but it was also somewhat conservative since actual production of nitrogen exceeded 85 million tons.[43]

CHAPTER FOUR

POPULATING THE EARTH: DIET, DOMESTICATION, AND DISEASE

Introduction

Farming and the sedentary life are associated with an increased rate of population growth. On the eve of agricultural invention, the estimated total size of the world population ranged from 100,000 to a few million people. Over the course of the entire Pleistocene Epoch, the human population grew insignificantly despite the fertility of the human female, the availability of sufficient food for small bands of hunter foraging populations, and the absence of infectious diseases. Such diseases would become commonplace once humans began the slow process of domesticating wild animals. So, what would explain this pattern of slow growth?

According to archeologists and evolutionary biologists, population growth was close to zero for thousands of years. On average, the number of births each year came close to matching the number of deaths. One rather conservative estimate of population growth suggests that the human population of several hundred thousand grew less than 0.01 percent per 1,000 individuals each year in the 30,000 years leading up to the transition to agriculture. This exceedingly slow growth estimate means that the global population would double only every 8,000–9,000 years.[1]

Another estimate, more precise in its calculations, allows 5,000 years for the maturation of an original pair to a stable population of 25 persons, 85,000 BP. With an annual growth rate of 1.000125 (0.0152 percent), which is very low by modern standards, a doubling of the world's population would occur every 4,561 years. As a result, the world population may have doubled during the very important developmental years of agricultural origins between 12,000 and 7,500 BP. Given the 85,000 years between the point of stabilization at 25 people and the end of the

Pleistocene, a world population of 8,605,565 at the beginning of the agricultural transition seems plausible.[2]

A Modern Demographic Scenario

Based on these calculations, a human population of almost nine million at the end of the Pleistocene would mean that probably no more than 50,000 people lived together anywhere in the world. These figures seem accurate, given the expansion into the Americas about 15,000 years ago, when the world's population was more than four million. According to Alfred Crosby, it would have taken only 400 males and females crossing the Bering Land Bridge 15,000 years ago and growing by 1.4 percent each generation to produce a modest population of 10 million North and South Americans by 1492 CE.[3]

In environmental terms, a human population of this size left a very small "ecological footprint," namely, the amount of natural resources humans use, despite the accumulating evidence of megafauna overkill by early hunters on the Eurasian and American landscapes. With the transition to agriculture, however, and the clearing of woodlands for cultivation, widespread environmental damage and ecological displacement became commonplace.

With its spread from Southwest Asia to Eurasia 12,000 BP, the population grew at 0.04 percent per 1,000 individuals each year. This represented a doubling of the population in less than 2,000 years or from several million to 250 million at year one on the Christian calendar.[4] Although no world census data exist for these historical periods, estimates suggest that about 150 years after Columbus reached the shores of Hispaniola in 1492 CE, the global population reached 500 million in 1650 CE. It took another 150 years for the population to reach one billion in 1800 CE. So, it took millions of years for the world's population to reach a population of one billion people. This very modest and slow growth contrasts significantly when compared with explosive modern population growth statistics.

It took only 123 years for the population to reach two billion, 33 years to reach three billion, 14 years to get to four billion, and another 13 years to reach five billion. In 1999, world population hit the six billion mark. The term "population bomb," popularized by the biologist Paul Ehrlich, doesn't begin to describe the impact of this explosion on the

Figure 4.1 Crowd in the Piazza del Campo, Siena, Italy before the running of the Palio horse race in which the city's numerous districts compete for victory. Its origins date back to the thirteenth century. *Source*: Permission granted by Peter Beaton.

ecology of the planet. Thinking about these changes in the size of the human population can become a daunting exercise. Simply put, during the past 650,000 years, humanity grew at the very slow rate of about 0.2 percent every thousand years. At the beginning of the Christian era, growth rates accelerated to about 0.04 percent each year and reached 0.08 percent in 1650 CE. They rose to 1 percent between World War I and World War II and reached 2.0 percent for a brief period after 1950 and now is 1.3 percent. These seemingly small percentage changes represent an explosive growth of humanity from about 2.0 percent growth every thousand years in the millennia before agriculture to 2.0 percent each year in the 1950s. This change represented a 1,000-fold increase in the human population.[5]

The total number will continue to rise to just under nine billion in 2050. Pessimistic predictions about a continuing "population bomb" that will further degrade the global ecosystem and sink humanity deeper into a morass may be overstated, however. Assuming we reach the predicted growth in 2050, a most recent worldwide demographic report by the United Nations no longer calculates the world's fertility

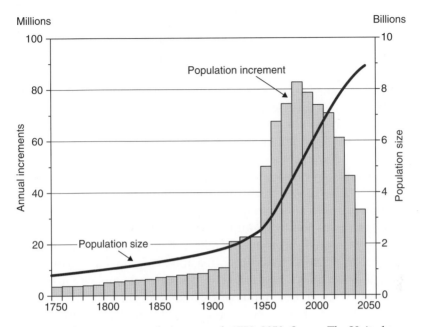

Figure 4.2 Long-term population growth 1750–2050. *Source*: The United Nations Population Division.

rate at an average of 2.1 children per household.[6] Parents have two children who eventually replace them. The 0.1 represents children who die before reaching maturity. At that growth rate, the world's population stabilizes over time. The new calculation based on trends in fertility today is assumed to be 1.85 and not 2.1 and the former figure is below the population replacement level. So, rather than a future global population explosion, the world will enter an extended period of depopulation.[7]

The Role of Disease in Calculating Population Size

Although the debate about the causes of slow population growth over the millennia has not been resolved, the biological and ecological factors involved offer a plausible explanation. As the size and density of human populations accelerated, the spread of infectious diseases increased. This happened because infectious microbes multiply rapidly, complete their life cycles quickly, and spread from one person to another. Breaking the chain of infection interrupts the rapid multiplication and maturation of microbes and their spread. Since the size and density of human populations are important variables to understanding the spread of disease, it is understandable why small, mobile bands of foragers made the task of transmission by parasites very difficult but, as we shall see, not impossible.

Zoonotic diseases such as rabies, anthrax, and ringworm can be transmitted from vertebrate animals to humans. With humans serving as the host, they posed threats to the health of other hunter-gatherers. As hunters, they came into direct contact with the flesh and blood of animal carcasses and by handling game and fur-bearing animals regularly, they exposed themselves to a host of potentially lethal infectious diseases. In addition to anthrax, the following infections sometimes resulted in fatalities: *toxoplasmosis*, a disease of mammals caused by a toxoplasma transmitted to humans via undercooked meat; *hemorrhagic fever*, a viral infection such as dengue or Ebola that results in fever, chills, and profuse internal bleeding from the capillaries, and *salmonella*, food poisoning caused by infection with organisms, usually characterized by gastrointestinal upset, diarrhea, fever, and occasionally death.

In addition to these harmful infections, butchering a kill often exposed hunters to the intestinal contents of the game and the very lethal *anaerobic bacteria*, including the carriers of gangrene, botulism, and tetanus.[8] Add to these diseases trichina worms (*trichinosis*) transmitted to humans from animals, *bubonic plague* carried by fleas, and *malaria, yellow fever*, and *West Nile fever*-bearing mosquitoes and you have a host of parasites that potentially compromised the health of a sparely populated and mobile human population.

Small mobile populations offer parasites limited opportunities to spread. "Parasites must complete their life cycles *repeatedly* and must spread from person to person (or they must spread repeatedly from the soil or animals to people) if they are to pose a threat to human populations; if the chain of reproduction and dissemination is ever completely broken, the disease disappears."[9] For hunter-gatherers such diseases have few opportunities to spread. They were subjected to parasites in the tropics where such organisms abound but as they migrated away from hot climates to cooler, drier environments such parasites were less likely to multiply.

Entering new environments, however, exposed *Homo sapiens* to new virgin soil parasites and resulted in episodes where infections spread. The activities of the hunt would also have exposed them to zoonotics. Low-mortality infections afflicted many, with most recovering except the most vulnerable, namely, the very young and the very old. Many parasites with their enormous ability to mutate would have traveled to the new climates with their human hosts and adapted with them.

In order to travel, these mutating micro-predators must have survived in low-population-density environments typical of hunter-gatherer communities. The parasites probably caused low-mortality diseases, since their human hosts would have developed immunity against a parasite's more lethal effects. With the rising and more humid temperatures of the last phases of the Pleistocene about 20,000 to 12,000 BP, vector-dependent parasites, meaning disease-bearing microorganisms that carry the infection from one host to another, including *anopheline* mosquitoes, would cause outbreaks of high-mortality malaria among humans. It afflicted women in their first pregnancy particularly hard, often causing the loss of the fetus and sometimes causing the death of the mother.

As migrating humans pushed into new frontiers, they participated in the megafaunal overkill and as a result reduced the number of animal hosts that parasites could invade. Humans then became a "mono-crop" for many parasites that adapted to their new human hosts. "Many of the most important human diseases, such as malaria, smallpox, epidemic typhus, and syphilis (congenital and epidemic), lacked alternative animal reservoirs. In turn, a human's highly efficient immune system learned to recognize and combat each new invader, but not without a cumulative toll."[10]

Slowly, they infected humans throughout the terminal Pleistocene and they played an important role in Holocene history of Southwest Asia, including the Fertile Crescent in Mesopotamia. The impact of these new density-dependent diseases would continue to be felt up to the present. At the beginning of the Holocene, world temperatures were probably slightly higher than they are today, intensifying the movement of micro-parasites from the tropical rainforests to the thinly settled, stable, and relatively healthy regions of Southwest Asia, the Mediterranean coast, and North Africa.

Remember that with a theoretical natural population increase of 1.000152 percent annually, the human population during the Pleistocene theoretically doubled every 4,561 years. If Pleistocene hunter-gatherers reproduced very slowly given the nature of a mobile life, then the presence of a high-morbidity but low-mortality disease would bring population growth close to zero. If one considers that a hunter-gatherer woman gave birth to her last child at age 30, then a disease with a mortality rate of only 0.44 percent would bring population growth to a standstill.[11] Slow-growing populations would become very vulnerable to diseases of only moderate mortality, mostly afflicting those of women of childbearing age and children.

The immune system of humans responds effectively to diseases that depend on dense populations. These include measles, mumps, influenza, the common cold, and tuberculosis, among others. Within a growing population, they trigger individual immune responses among mature adults, thereby turning wide epidemics into age-specific diseases among the young. As the population of young people grew, these density-dependent diseases grew accordingly. Many other common infections such as tetanus or botulism will not progress in the same manner. "Similarly, rare epidemics of irregular occurrence, no matter how catastrophic in short-term impact, have little lasting effect on natural increase, and, most importantly, recurrent diseases of severe mortality tend to be self-regulating and leave little long-term imprint except in the immune system."[12]

When the mortality rate from infectious diseases rises to 2 percent annually, the afflicted population responds by exerting selective pressure to save and promote those genetic traits that improve a person's immunity against the disease. With births occurring during these disease episodes, the number with immunity grows faster than those who acquired immunity through exposure. The space between outbreaks of the disease slows and may stop as the entire population becomes genetically protected. High-mortality diseases result in strong human immune responses.

In hunter-gatherer communities, no such high-impact diseases are necessary to slow population growth. Infections with a low mortality of about 1 percent or less will not trigger an immune response, where natural selection will save or promote advantageous genes. So, infections can continue to revisit stable populations, retarding population growth. Cultural traits may have developed in the absence of genetic responses to low-mortality diseases. Using fire to cook meat and fish thoroughly protected the practitioners from some zoonotic diseases, while hand-washing and other forms of personal hygiene protected them from others. For example, prohibitions against eating pork, a source of infectious disease, have their origins in culture, religion, and health.[13]

The Impact of Migration and Settlement on Global Population Growth

Some researchers make the argument that the onset of a sedentary life is part of an evolutionary process, with mobility among foragers decreasing, population increasing, and food gathering and/or cultivation becoming more specialized. "If a shift occurred then a long sustained pattern of Pleistocene mobility and population regulation ended with the emergence of large, less mobile, and more complex communities."[14] However, becoming less mobile may have begun much earlier as humans mastered fire at least 400,000 BP.

With the completion of this first wave of human migration, hunter-gatherers used fire to improve the yield from the hunt and learned about the fertilizing quality of an ash-littered soil. This learned behavior would serve them well thousands of years into the future and drastically change the relationship of humans with the natural world. By burning grass and underbrush during periods of drought, hunters improved the forage for animals and opened more unobstructed sight lines for the hunt. By burning the groundcover, fresh ash fertilized the soil for new growth that cultivators used as forage for animals or as foods for hunters.

The complexity of this process and the central role of controlled fire undoubtedly changed local environments and taught mobile bands to hunt within a clearly circumscribed area whose landscapes they regularly altered by burning. Settlement and mobility competed with each other over thousands of years, with agriculture eventually displacing hunting and gathering. Early farmers discovered how to make fields on forested land by using a stone ax to girdle trees, slashing their bark all the way around, thus interrupting the flow of sap to upper branches, resulting in the death of the trees.

Then, after crops of grain were planted in the leaf mold of the forest floor for a few years, the dried-out skeletons of the dead trees could be set on fire and their ashes used to fertilize the depleted soil. After a few more years, however, wind-blown weeds invaded such fields, and, as long as forested land remained within their reach, slash-and-burn cultivators found it easier to repeat the cycle by opening up a new field somewhere else than it was to stay put and allow the weeds to diminish the harvest.

Abandoned clearings then went back to forest as nature took its course, and invasive weeds gave way in natural succession to taller bushes and then to trees. "As long as they remained few enough, slash and burn farmers could therefore establish something approximating a steady relation to nature by returning to such plots at suitably long intervals."[15] Modest changes in the size of a farming population required expansion onto new lands and the temporary departure from old fields, returning to them only after environmental renewal replenished the soil's nutrients.

An increase in the number of people practicing farming and living in a carefully defined area was a consequence of abandoning the life of hunting and gathering. By embracing agriculture an increase in population followed, reflecting a more sedentary life and a quickened fertility cycle. One estimate places the movement of early farmers from Southwest Asia to Europe during the Neolithic at one kilometer annually.[16] With each advance of farmers and the spread of agriculture, deforestation eliminated the habitat familiar to hunter-gatherers, wildlife, and wild plants. The earth became transformed with each migration. Deforestation compromised watersheds, rivers, and streams with soil runoff and exposed the land to shifting climate conditions and changes in the local microclimates.

The Role of Nutrition in Early Population Growth

Although hunter-gatherers ate a more balanced diet than early farmers, were somewhat larger in stature, and suffered from less tooth decay, the resources available to them in the natural world limited the size of their population and the rate of increase. A population of foragers can use only a habitat within a day's walk. They must be able to leave and return to their camp daily. The availability of food depends on the ecology of the habitat and its accessibility. If these functions cannot be successfully completed in a single day, then the foragers must move to another area. So, plants and animals and the amount of precipitation that they require in order to survive are essential ingredients for a population of hunter-gatherers. The availability of plants and animals placed limits on the ability of foragers to feed their families. To sustain this way of life, population densities seldom exceeded one person per square kilometer.[17]

Population densities grew slowly over the millennia during the transition to agriculture. Changes in cultivation from "slash and burn" to tri-annual rotations of crops, the selection of better seeds, the domestication of plants and animals, and the eventual harnessing of animal and water power to increase supplies of food resulted in densities of 40–60 persons per square kilometer in European countries by the mid-eighteenth century. In some regions this change in population density was 100 times greater than that of hunter-gatherer densities.[18] The cultivation of highly nutritional and easily stored grains such as wheat, barley, millet, corn, and rice expanded food supplies and reduced the nutritional problems created by diminishing yields during periods of ecological stress. With these changes in productivity, one would assume that the health of farmers improved and their mortality declined. Was this the case?

The limited evidence suggests the following. A maize-based diet in Mesoamerica during the early Neolithic period may have caused increased dental decay but in coastal and inland Europe, with a protein-rich diet in shellfish (oysters, mussels, and clams), venison, wild boar and rabbits, no evidence exists of rapid dental pathology. The general reduction in the size of adult second molars began long before the transition to agriculture and may be explained best by the transition from breastfeeding to weaning. Second molars begin to develop around 36 months, after the age of weaning in most modern societies. Among foraging populations, breastfeeding may have been extended into the second year of a child's life and the spacing between children may have been approximately 36 months. Three-year-old children may have suffered from short-term malnutrition caused by the addition of a newborn. This conclusion does not reflect a general dietary decline as a result of the Neolithic transition but refers to a specific nutritional shortcoming. In Europe, the evidence of a greatly impoverished diet as a result of the transition to agriculture is fragmentary.[19]

Seemingly, fertility among farmers in settled communities exceeded the rates of fertility among highly mobile bands of hunter-gatherers.

"In human populations, numerous factors – including age of marriage, the length of a woman's reproductive period, birth spacing, coital frequency, and the importance of contraception can affect the number of births each female has during her lifetime."[20] Among settled populations, as the period of breastfeeding infants diminishes, the chances of conception increase. A rise in the number of live births seems to have been the result.

The absence of proper weaning foods may have extended nursing among hunter-gatherer mothers. Difficult-to-digest staples such as tubers, nuts, and other wild plants would not have provided for the nutritional needs of small infants who eat relatively small amounts of food many times each day. The high-bulk diets of adults with relatively few calories compared to other nutrients may not have been suitable for infants. Diets low in fat content protected adults from degenerative diseases but probably did not satisfy the nutritional needs of infants. A life on the move made it difficult to care for a number of infants at any given time. Combined with an extended period of nursing the young from two to six years and intensive physical exercise by hunter-gatherer women, their fertility diminished temporarily.

In contrast, the high-carbohydrate diet of early farmers of the Levant in Southwest Asia, with ample supplies of easily digestible wheat and barley, would provide growing infants with alternative sources of nutrition to mother's milk. To supplement this nutrition-rich diet, infants in early farming communities would have milk from goats, sheep, and cows to accelerate the weaning process. Improved nutrition promoted human cellular growth. As a result, natural contraception, a byproduct of mothers nursing their infants, ended incidentally with the increasing availability of nutritious cultivated crops.

Although the transition took millennia to unfold, with hunter-gatherer and early farming practices coexisting for long periods, farming was a labor-intensive activity that took its toll on the population. What we now know is that in sedentary agricultural communities, mortality and fertility rates rose but with a slightly higher birth rate, thereby explaining the growth in population. What explains the higher death rates among early farmers when compared to hunter-gatherers? A decline in the wide variety of nutritious wild nuts, fruits, and tubers, as farmers cleaned the land for a few specific domesticated crops, may have led temporarily to a decline in overall nutrition.

The Role of Animal Domestication in the Spread of Infectious Disease

A second even more convincing explanation is suggested by the conditions of a stable and settled population and by the spread of infectious diseases. These biological invasions occurred as farmers brought animals closer for domestication and food. Since we know that many human diseases have their origins in the close contact between people and domesticated animals, their association retarded the natural growth of the human population. Also, settled hamlets and villages concentrated people in the vicinity of their own wastes, potentially fouling their land, water, and food. They became environments for the incubation, spread, and survival of microbes and infectious diseases, then unknown or rare among foraging populations.

The transition from a small, mobile, isolated life to the more sedentary agricultural society probably limited human exposure to a number of zoonotic diseases. Settlement also provided more opportunities to care for the sick and return them to health. Sedentary societies increased their trade with neighbors both near and far, however, exposing themselves to local micro-variants of common infectious diseases. Microorganisms located in one's intestines, however, may protect persons from other local invasive potentially pathogenic organisms. In the process of fighting off local pathogens, these microorganisms protect persons from outside invasions.[21]

The domestication of animals that proceeded along with the cultivation and domestication of wild plants probably increased the vulnerability of early farmers to infectious diseases. Close encounters with wild and domestic animals place humans at risk from infectious microbes carried in animal fluids and wastes. Dogs can be carriers of rabies and to a lesser degree, the dreaded tetanus, carried primarily by horses. Tetanus can also exist in grazed or farmed soil. Originally, smallpox and tuberculosis were thought to have originated in cattle but now we know that camels are the source of smallpox infections and tuberculosis probably has multiple animal sources, including buffalo, wild bird populations, and others. Diseases in cattle are the sources for measles and diphtheria. Influenza comes from avians, pigs, and chickens, while the common cold began with human contact with horses. Most human respiratory infections probably began in this way.[22]

93

Growing human population density increased the survival rate of microorganisms needing human hosts in which to generate more offspring. Sparsely populated areas decreased and in some cases eliminated the ability of parasites to infect one person after another, regardless of whether transmitting the infection occurred from person to person or from the parasite through the food chain, soil, the air, or water. Human population density and the physical closeness of domesticated animals provided the environment for many of our infectious diseases. Mobility, group size, isolation, and the influence of climate are important factors in understanding the transmission of disease. Fewer pathogens and fewer vectors can survive in colder and drier climates. Parasites incubated in human waste don't normally survive in desert conditions, nor do vector-borne diseases such as malaria carried by mosquitoes. As *Homo erectus*, a million or more years ago, left tropical Africa and as anatomically modern humans migrated more than 100,000 years ago for suitable cooler and drier habitats, their health improved and their mortality rates probably declined.

The role of climate in the dissemination of disease microbes cannot be minimized. Each microbe has environmental limits outside of which it cannot survive. The role of climate is apparent in the spread of infection by aerial transmission. In this regard, the common cold, influenza, measles, pneumonia, and smallpox spread during dry, cool seasons of the year. Vector-borne diseases such as malaria, yellow fever, and sleeping sickness require tropical temperatures. These seasonal climate effects may be accompanied by human activities such as holidays and festivals that draw large numbers of people, facilitating the spread of infection. Other environmental effects caused by human intervention include forest clearance that provided open space for grazing and farming which encouraged the reproduction of mosquitoes that spread malaria. Changes in cattle grazing also influence the spread of infection. Cattle herds may offer mosquitoes another source of blood meals, thereby encouraging the spread of malaria.

The conclusion that those members of the world's population who made the transition from foraging to farming suffered from poor nutrition, increased mortality and morbidity due to infectious diseases, reduced stature, and increased dental pathology may be generally correct but with some important exceptions. The data supporting this generalization come from specific regions of the world, primarily the North American Midwest and Sudanese Nubia, not from the Eurasian landmass, where farmers from Southwest Asia expanded into Europe replacing foragers in their path. Were these migrating farmers weakened by malnutrition and infections caused by settlement and overpopulation or were they healthy and fertile, thus able to "conquer" a continent rapidly?

In short, why would a population incur a serious health risk by making the transition from hunting to farming? The simple answer is that they wouldn't unless they were forced into it without any other reasonable choices. The forcing mechanism may have been pressure exerted by micro-parasites expanding their universe as a result of the global climate warming during the terminal Pleistocene. As most evolutionary biologists would argue, no species would select a genetic survival strategy in which mortality increased and life expectancy declined, endangering the viability of the species. "Biological evolution is powered by the selection of strategies for improved success in breeding; strategies that threaten breeding will not be selected even if, in the long term, such strategies would improve efficiency. As evolutionary biologists keep insisting, selection is blind. It does not and cannot 'plan'; it merely responds to immediate breeding success."[23]

Once the food quest changed from foraging to the intensive cultivation of crops and the domestication of animals, reproductive effort had to be constantly improved to ward off infectious micro-predators and to stay somewhat ahead of them. Human migrations, as groups intermingled peacefully or invaded and conquered others, enhanced selective pressure on *Homo sapiens*. The intensive cultivation of crops that included selecting plants with high yields and the domestication of goats, sheep, and cattle and their selective breeding to improve the quality of their offspring were part of the reproductive effort to keep ahead of diseases.

So many new viruses and new strains of old viruses spread through settlement communities. For example, respiratory infections spread rapidly as people raised fowl and pigs for human consumption. Birds shed viruses in their waste; pigs eating from the ground ingest the viruses. Since pigs can carry human as well as bird viruses, they create a biological environment in which genes can be exchanged. This exchange led to the creation of new strains of old viruses and new viruses that infect humans. Cross-species infection is particularly destructive today in villages and cities where human and animal densities reach high levels. In many developing parts of the world, millions of people live in quarters with less than 40 square feet per person. By contrast, the modern college dormitory room has about 150 square feet.[24] As a result, infectious diseases evolved, mutated repeatedly over the millennia, and spread rapidly to humans.

As these populations came into close contact with others through migration and exploration, they spread infections to populations without acquired immunity. The use of the term "the great dying" signified the near-collapse of the Amerindian population in the early years of the sixteenth century by diseases brought from Europe. "Of the approximately 54 million people living in the Americas in 1429 CE only 5.6 million remained in 1650 CE. This number represents only 11 percent of the original figure; a population crash of striking proportions crushed once thriving societies. Population declines were the highest in Mesoamerica and the Caribbean and lower in the less densely populated regions of North America. With repopulation from Europe and regeneration of Amerindian people by 1750 CE, the total population of the Americas may have been only 30 percent of its 1492 estimates." Small demographic growth or no growth whatsoever became a worldwide phenomenon until the fifteenth century. Then, global populations began their steady rise, as the result of an explosion in the globalization of food crops, the migration of Europeans, and the forced migration of Africans through enslavement.

Despite the invention of agriculture and our image of a steady supply of food for human consumption to fuel the growth of the human population, the long-term history of humanity was one of misery caused by the pressure of population against available resources. Chronic food shortages kept death rates very high for very long periods. Combined with infectious disease, it acted synergistically with long-term malnutrition and created the conditions for high death rates. Without the latter, humanity might have experienced a many-millennia decline in mortality.

Nutrition, Climate Change, and Population

Nutrition and diet are not the same, however. Individuals and groups may have sufficient food and still suffer nutritional losses. A person's level of physical activity, the climate of the region in which the individual lives and works, and exposure to different diseases determine his or her nutritional needs. One example among the literally thousands that make this conclusion meaningful is the following. Low levels of iron consumption increase one's susceptibility to hookworm, a disease endemic to people today in underdeveloped parts of the world. Agriculture may have provided for the nutritional needs of the population but not if the claims on that increased food supply included more arduous work and extended exposure to infection. These considerations complicate our understanding of the benefits gained by humanity for making this important historical transition from hunting and gathering to agriculture.

As noted earlier, the average body stature and weight of humans declined with the long-term pattern of malnutrition after the transition to agriculture and the emergence of many new infectious diseases. With this accumulated information, the evidence points out that humans were severely stunted as measured by the stature of Europeans as recently as the eighteenth century. This long-term pattern of stunted growth can only be accounted for by a caloric intake of 1,750 calories per person in that century, similar to the current intake of persons in impoverished nations today.

Across Europe and Asia, populations began a steady upward recovery from the Black Death in the fourteenth century. In 1300 CE, before the outbreak of this Eurasian epidemic, the world's population had reached 360 million. By 1750 CE, the world's population had not only recovered from the plague but also doubled in size to some 750 million people.[26] A majority, however, was unable to work long hours or engage in meaningful activity. Many of the worst-case scenarios eliminated more than 20 percent of the population from meaningful work because of their poor diets and low body weight. In fact, they were wasted, weighing 25–30 percent less than those with higher caloric diets. In part, this explains why as many as 20 percent of the population in France before the revolution in 1789 CE were beggars.[27]

The decline in populations between 1348 and 1485 CE was associated with repeated outbreaks of bubonic plague. Estimates of the death rates in Europe alone, during this century, range from one-fourth to one-third of the population. The global estimates of death rates from plague remain elusive. What we do know is that plague is spread by migrating flea-infested rats that come into close contact with humans. During times of food shortages and famine, rats migrated in search of food. Historians searching for the causes of famine have found striking similarities between food shortages and changes in the global climate in the temperate latitudes over periods much longer than the centuries described here. People living from 1348 and 1485 CE experienced climatic cooling. London's Bills of Mortality, published weekly, listed more than 60,000 deaths caused by the plague that enveloped the city from April through October 1665.

So, a relationship between population growth and decline is associated sometimes with climate change and fluctuating food supplies. Given a human population measured in millions and not billions in the early centuries of the last millennium, global or hemispheric climate change occurred without human action. Long-term variations in solar activity, a contributing factor in all climate change, were the primary cause for the long-term fluctuations discussed here. Comparing evidence from mineral deposits in New Zealand stalagmites, carbon-14 deposits and concentrations in ancient trees, and historical documents describing changing weather patterns, mortality records, agricultural productivity, and others, the middle latitude temperatures show remarkable consistency.

People living during the Medieval Warm Period (800–1200 CE) witnessed an increasing global population. Better tools and replacing oxen with horses whose pulling power was enhanced by new harnesses improved agricultural productivity. Using legumes and fertilizers reduced the depletion of vital soil nutrients. Increased production led to the growth of towns and an expansion of regional trade. Declining temperatures, reaching a low from 1590 to 1670 during the Little Ice Age (1300–1850 CE), led to food shortages and rising prices. From 1500 to 1650 CE grain prices in the Ottoman Empire, China, and British rose by as much as 500 percent.[28] Grain yields fell, death rates increased, and fertility rates and life expectancy at birth declined. A warmer period followed from 1670 to 1800 CE, as the Little Ice Age retreated, ushering in longer growing seasons, an increase in foodstuffs, and declining prices.

The characteristics of these rapid fluctuations are worth noting for their historical and current implications. The climate changed rapidly as well as gradually in past time. It profoundly affected the lives of people who experienced these rapid fluctuations and had local, regional, and global effects. Cooler temperatures showed greater variability in week-to-week and year-to-year weather. And most importantly, a cold and variable temperature made food production precarious. In these agrarian societies, a rise or fall in agricultural production affected economic wellbeing. In long cold periods, the amount of land for cultivation is reduced as land at higher elevations becomes unavailable. A 2 °F change will shorten the growing season by three to four weeks. Famines

Figure 4.3 *Hunters in the Snow* (1565) by the Dutch painter, Pieter Brueghel. The painting is used as supporting evidence of the severe winters during the Little Ice Age. *Source*: Kunsthistorisches Museum, Vienna, Austria.

are a logical outcome of declining food supplies and, for the most vulnerable in society, starvation and increased susceptibility to infectious disease result in greater mortality.[29]

Astronomers, using powerful telescopes invented in 1609 CE, began to make systematic observations of the Sun's activity and observed first the causes for these changes in climate. Initially, they witnessed significant sunspot activity, namely the discharge of solar energy that some astronomers believed affected conditions on Earth. Within decades, however, these keen observers recorded a decline in sunspot activity to the point where fewer were recorded between 1645 and 1715 CE than appear now in one year. With solar energy reaching the Earth decreasing, at least 12 major volcanoes erupted between 1638 and 1643 CE, creating a cloud of dust in the atmosphere that blocked sunlight from reaching the Earth's surface.

Dendrochronologists measure the impact of these solar and terrestrial events on the climate by studying tree ring growth. Each year, a tree grows a ring, the size of which is determined by favorable or unfavorable growth conditions for plants. For example, moderate temperatures and average seasonal rainfall promote growth, while unseasonable long, cold and dry winters stunt growth. Frost rings appear when summer temperatures fall below the freezing point. The dendrochronological record for surviving seventeenth-century trees reveals that three of the coldest summers in the past 600 years occurred in the 1640s. The Little Ice Age (1300–1850) had overwhelmed the middle latitudes of the northern hemisphere.[30]

Crop yields declined from 30 to 50 percent in the 1640s, with slumps in cereal production recorded in Hungary, Poland, and Spain. Harvesting grapes in France occurred two weeks later than normal. The Thames River in London remained frozen throughout the winter, while mountain peaks held their snow throughout the year. Polar ice interfered with sea and channel traffic and advancing glaciers removed valuable fields, farms, and villages from production. Under temperate climatic conditions, Western European farmers could expect a growing season of as many as nine months while further to the northeast in Imperial Russia a normal season ranged from four to six months, north to south. A near-doubling of the population from 1450 to 1600 CE meant that the impact of this ice age would exacerbate living conditions for more people than at any time in human history.

With poor harvests, rising food prices, and an overall slowdown in economic activity depressing wages, this larger population faced a catastrophe. When a grain harvest failed, the staple in the diet of most poor families, hunger was followed by a general famine. Poor families spent between two-thirds and three-quarters of their incomes on bread, the cheapest source of calories. Ten pounds of bread cost the same as a pound of meat. A nutritionally weakened population became quickly vulnerable to infectious diseases, typhus, typhoid, dysentery, and bubonic plague. In France, one million persons died in the plague epidemic of 1628–1631 CE. In Venice, 40,000 died in 1630 CE, while in the kingdom of Naples, some of its cities and towns lost one-half of their populations.[31]

By the sixteenth century, population growth accelerated in Southwest Asia, with estimates ranging from a 50 to 70 percent increase between 1500 and 1570 CE. Some towns experienced growth rates of 200 percent or more. Istanbul, with a population of about 100,000 in 1520 CE, became a sprawling metropolis of 700,000 in 1600 CE. China's population swelled from 45 million in the latter decades of the 1300s to 150 million by 1600 CE. The major urban centers of China, namely Soochow, Nanjing, and Beijing, grew significantly while smaller market towns experienced explosive growth. Japan, one of the world's most densely populated countries, saw its population grow from 7 million in 1200 CE to 12 million in 1600 CE and take off to 31 million in 1720 CE. England's population of 2 million in 1520 CE became more than 5 million in 1640 CE, with London's growth leaping from 50,000 in 1500 CE to 400,000 in 1650 CE.[32] Market towns such as Norwich, Worcester, Bury St Edmunds, and many much smaller places doubled and tripled in size.[33]

Associating long-term solar variations with carbon-14 concentrations in ancient tree rings, the population growth in Western Europe and China appears to be very similar. Population grew during warming periods and either remained stable or declined during periods of global climate cooling. The global population explosion began with this warming trend and accelerated after 1750 CE. Changes in the global climate served as a trigger for increased crop yields providing more food for each person. Larger and healthier herds of cattle produced a larger supply of meat; the milk yield of cows improved and a myriad of factors increased fertility and lowered infant mortality.

Industrialization, the topic of Chapter 7, was under way and it will become apparent that no single factor, neither the substitution of steam power for water, nor the replacement of woolens and linen with cotton, nor some important inventions including machine looms and canning of foods, fully captures the essence of industrialization. As the economist Joel Mokyr has pointed out, "The Industrial Revolution was more than a conversion of organic to inorganic materials or the adoption of machinery. It was, quite simply, the adoption of a host of techniques that worked better, cheaper, and faster."[34]

As the evidence suggests, however, climate change triggered rising per capita incomes in some European countries and a population explosion throughout the continent. Economic growth, with rising incomes and rising standards of living, triggered a rise in the numbers of people marrying. In England from 1600–1649 CE, 34 percent of all women married were aged 22 and under but in 1750–1799 CE this figure rose to 50 percent. Favorable economic times and good weather changed the behavior and the fortunes of a fecund population.[35] Three factors explain this important demographic transition. Mortality, fertility, and migration directly influenced population levels, when *Homo sapiens* came to dominate the planet. Add the important variable of economic development in the form of extensive agriculture and rapid industrialization and you have most of the significant factors explaining the population explosion of the past centuries.

By current standards, the majority of the world's population remained undernourished as recently as the eighteenth century, even among persons in the top half of the income distribution. Improved nutrition was brought about by increased agricultural yields that made more food available for consumption. Improvements in farm technology reduced the number of workers needed to produce greater yields. More people then became available for work in towns and growing cities and fewer people with poor diets remained outside the mainstream of productive society. Excellent pre-industrial harvests in the 1730s in Britain coupled with an unsurpassed birth rate had the effect of expanding the market for goods and services. A well-nourished population would become an important variable in the making of an industrial world. Populations had surged before in the pre-industrial world but their growth had been eclipsed by an agricultural society's inability to increase production to meet demand. Malnutrition made populations susceptible to disease, resulting in a demographic collapse. A truly demographic and epidemiological transition began to unfold in the eighteenth century in British and paved the way for economic acceleration.

A population boom began in the 1730s and accelerated as boomers matured, married, and raised families. Assisted by a gradually warming climate, growing food supplies would add to the nutrition of the larger population. Between 1660 and 1740 CE, food consumption per person increased by as much as 50 percent. Small changes in temperature extended the growing season, increasing the length of time that grazing animals spent in pastures and resulting in greater meat and milk production. The warm and dry years of the 1730s and 1740s led to improved wheat yields and may have promoted human health. Damp houses create the conditions for respiratory illness, especially among infants and young children. The greater availability of food and declining costs extended beyond British as warming temperatures made more food available in much of continental Europe, Asia, and the Americas and the mix of wheat and legume consumption increased both caloric and protein intake.[36]

Increased pre-marital pregnancies, as high as 35 percent at the end of the eighteenth century, early marriages, increasing birth rates, and declining infant mortality created this population boom. In order to absorb this growing population into the economy, British needed to sustain its development in agriculture and industry, add to its growing urban sector of cities and towns, build its physical and human capital, and gain access to continental and overseas markets. To keep up with the population explosion, it discontinued exporting food. Its beef imports from Ireland tripled, with butter and pork increasing substantially in the 1750s and 1760s. Prices soared, causing growth to slow and food consumption to drop by as much as 20 percent. Previous nutritional levels were high enough, however, for the population to weather the decline without experiencing famine.[37]

The intertwined relationship between population growth and industrialization became a model for those who followed in the next century. Population density meant that the distance between buyers and sellers reduced transportation and transaction costs. By 1750 CE, one-fifth of the English population lived in towns, a high level of urbanization for that century. Fifty percent of the population increases since 1600 CE came from the cities and towns, with the spread of the factory system providing a ready supply of available workers – men, women, and children. Although their wages were low by modern standards, laborers nonetheless demanded industrial products.

The last time that population surged in the sixteenth century with an urban population of 5 percent, it suffered from the inability of agriculture to grow sufficiently to accommodate the boom. In the 1700s, however, urbanization at 20 percent was accompanied by a growing commercial presence that brought British and Europe closer to distant agricultural regions. These provided nutrients during periods of population growth.[38] Good harvests and rising nutrition contributed to a population boom that provided the labor pool for industrialization.

Improvements in housing and sanitation practices in Europe and the United States contributed to lower mortality rates before the development of vaccines and antibiotics. Better hygiene and greater care with the preparation of food kept infectious microbes at bay. Life expectancy rose by 30 years among members of the top half of the income distribution and by as much as 50 years among the lower half. The world's population began its steady rise from 1.6 billion in 1900 to 2.5 billion in 1950 to 6.1 billion at the end of the twentieth century.[39]

A Population Bomb or Not?

As the world's population continues to rise, alarming claims about an eventual population crash appear, echoing the predictions made by the British economist Thomas Malthus in his famous *Essay on the Principle of Population* published in 1798 CE.[40] As the world then approached one billion people, Malthus warned of a disaster caused by growing population and insufficient supplies of food. Although his assessment proved to be wrong at that time, fears about a pending demographic collapse continue, despite a population that now exceeds six billion and food production and consumption per person much higher now than ever before. Average life spans are lengthening and general living standards in many parts of the world are improving. With close to a billion people added to the world's population in the past decade, fears persist about a coming demographic crash.

Since about 90 percent of the increased population growth takes place in the developing world, especially in Africa, with more than a 3.0 percent annual increase, and in Asia, with the largest absolute increase, worries remain about an imbalance in global population, with the poorest countries growing the fastest. With populations in Africa and Asia rising from almost 64 percent of the world's population in 1950 to more than 70 percent in 2000, with an estimated increase to 78.5 percent by 2050, developed countries with falling birth rates fear becoming overwhelmed by immigration. According to Nobel Prize-winning economist Amartya Sen, the fear in the developed countries of becoming overrun by impoverished people of color is an overreaction, for the following reasons.[41]

First, the developing world is currently experiencing a demographic transformation not unlike the one experienced by Europe and the United States during periods of expanding industrialization in the eighteenth and nineteenth centuries. While Asia and Africa grew by 4.0 percent or less each decade during the nineteenth century, Europe and North America grew by about 10 percent each decade. Despite this extraordinary growth, the share of the world's population in Africa and Asia was then about 78.4 percent. Today, its share is 71.2 percent, much lower than it was in either 1650 or 1750. The United Nation's prediction is that Africa and Asia's share of the world population will rise again to 78.5 percent in 2050, a figure that approximates their share before

Figure 4.4 Waterside marketplace in Monrovia, the capital city of Liberia, on Africa's northwestern coast. *Source*: Michael Waite, Friends of Liberia.

industrialization in Europe and the United States.[42] So, the idea that the world's population has reached a point of imbalance ignores these historical trends.

A second concern articulated by a fear of an impending population explosion involves the availability of food. Will population growth outstrip our ability to feed more people? Despite episodic shortages of food in war-torn areas or in disaster zones causing millions to go hungry, the world's production of food is growing at a faster rate than population. This conclusion applies more dramatically to the developing world than to the richer countries, in terms of per capita income. Globally, food production rose 3.0 percent per capita, while it increased by 2.0 percent in Europe and fell by nearly 5.0 percent in North America. In contrast, food production per capita jumped by 22 percent in Asia and 23 percent and 39 percent in India and China, respectively. Only in sub-Saharan Africa, where food production fell by 6.0 percent, have famine and despair become a part of a general economic malaise caused by wars, political repression, and chaos.[43] With the exception of Africa, pessimistic predictions about a food crisis brought about by a population bomb seem unwarranted.

Along with the per capita increase in the availability of food has been a decrease in price, a price cheaper in constant dollars than at any time since Malthus wrote in the eighteenth century. During the 1990s, prices for basic foods such as rice, wheat, sorghum, and corn dropped by 38 percent. As a result, farmers have curtailed food production in order to prevent further price drops. If population growth were catching up with food supplies, then price incentives to produce more food would come into play. Currently, no such *push* in terms of a population bomb or *pull* in terms of the need to produce more food exists. Recent spikes in food prices, described in the Epilogue, may change this push/pull ratio, however.

Neither population growth alone nor increasing food supplies can explain away the persistence of poverty, overcrowding, and large urban slums in many of the developing world's cities. Old urban slums in the developed world and the persistent pockets of rural poverty almost everywhere remain a global problem. In the developing world, the slums of Calcutta and Mumbai, India, grow faster than those of Karachi and Islamabad, Pakistan, despite Pakistan's population annual growth rate of 3.1 percent and India's 2.1 percent. The same counterintuitive conclusion is true of the faster-growing slum conditions in Mexico City when compared to San Jose, Costa Rica, with an annual growth rate of 2.8 percent compared to Mexico's 2.0 percent rate. This same pattern exists when comparing the faster rate of growth in Indonesian slums when compared to Turkey and New York City's Harlem district when compared to Singapore. Economic and political processes other than population data and availability of food must be considered when explaining the conditions of a billion or more of the world's poor. Population density may be a contributing factor but not the cause.

In the way in which population and food are co-joined, the effects of population growth on the environment are seldom addressed with the fervor that Malthus inspired. An environment damaged by population pressures using more of the world's natural resources certainly affects our quality of life. On a per capita basis, however, countries in the developing world consume less food, fuel, and other resources than individuals in the developed world. Although a population of billions

Figure 4.5 Winter Crowd in Shibuya, a fashionable shopping and entertainment district in Tokyo, Japan. *Source*: Jean-Francois Berthet.

in the developing world may not adversely affect the global environment in the short run, their impact on the local and regional environment is another matter entirely. Unregulated local industries continue to pollute the soil, water, and air, while the depletion of local wood reserves for fuel degrades the forests.

Comparatively speaking, however, one additional American, Western European, and Japanese citizen consumes 32 times more resources than do dozens of citizens of the developing world and as a result typically contribute much more to global climate change.[44] As these dozens grow richer and consume much more, mimicking the behavior of consumers in wealthy countries, their long-term impact on the environment can be large. Their threat to the environment will become the same as the threat currently posed by the developed nations.

CHAPTER FIVE

THE MAKING OF AN URBAN WORLD

Introduction

In many parts of the world, the city is home to the majority of the population. For example, in the Americas as well as in Europe, 75 percent or more of the population now live in cities, while in the most populated regions of the world, namely Asia and Africa, no more than 38 percent of the population is urban. Even where the rural population still represents the majority on these latter two continents, by adding one million people each year the growth of cities will eclipse the rural majority by the middle of this century, if not sooner. Both continents are expected to grow at much faster rates than either Europe or the Americas over the next 30 years, with as many as 54 percent of the population becoming urban. The rate of urbanization and population growth there will far exceed that of the other continents.[1]

In every urban region, including those that have reached maturity, namely in areas where the growth rates have slowed considerably, cities contend with incredibly complex issues of providing for the nutrient balance of their residents. In effect, this means providing for their basic human needs for food, fuel, clothing, durable goods, transportation, commerce, and production. All of these nutrients create wastes, so each city must contend with methods and means of disposal. To paraphrase the urban and environmental historian Joel Tarr, each city, whether it is modern or modernizing, searches for the ultimate sink.[2]

For example, water for a city, either in the form of precipitation or as subsurface water used for human, residential, commercial, or production processes, ultimately becomes wastewater or waterborne sewage. Its treatment and disposal will affect the health of the urban hydrosphere and its residents. The magnitude of the sewage disposal problem

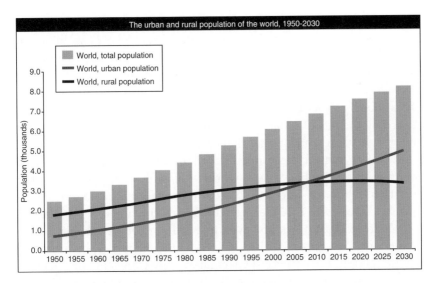

Figure 5.1 The world's urban and rural population. *Source*: Permission granted by The United Nations Population Division.

has literally overwhelmed the capacity of many cities in history to protect their residents from infectious diseases of epidemic proportions.

The wastes from other nutrients create their own urban complexities. Wood for fuel and construction depletes the surrounding countryside of its protective canopy, and leads to soil erosion, higher wind velocities, and dust and airborne debris. Burning wood and coal exacerbated this condition by polluting the air and the soil with contaminated atmospheric fallout.

A similar effect occurs when the built environment of cities, its contents and support systems, undermines the nutrient balance of the urban biosphere. The biodiversity and cycle of plant life provide biomass to replenish the soil's nutrients. The excavation for building sites removes sand and gravel that serve as a filter for natural water flows. Similarly, landfills made up of imported materials alter the ecosystems of these built environments. In addition to the erosion of topsoil, the fabric of the built environment includes surfaces made of composite materials (e.g., asphalt and concrete) that lead to storm water runoff and a decline in urban subsurface water tables. All aspects of the land's biosphere are transformed by urbanization and population concentrations.

To what extent are these processes slowing down or speeding up? The world's population will rise above its present level of six billion to an estimated nine billion in 2050. Currently, the world's urban population stands at 2.86 billion people and is projected to increase to almost five billion in 30 years. At the same time, the world's rural population will probably grow from 3.19 billion to about 3.29 billion people. So, regarding the rate of urban growth, governments, urban planners, and the citizenry face a staggering challenge in many parts of the world.

The effect of this concentrated population will result in an absolute increase in the number of cities worldwide, with some cities achieving the status of "mega-cities." To place this transformation in global perspective, the world's 200 largest cities each contained about 200,000 people in 1800. By 1990, however, five million people inhabited each of the world's largest 200 cities. In 1800, only Beijing, China held one million people. One hundred years later, there were still only 16 cities with a million or more inhabitants. By 1950, the number had risen to 86, and 50 years later in 2000, the number had reached 400. By 2015, estimates put the number at 550 and rising! Most of the new entrants to this exclusive club of cities will be located in Asia, Africa, and Latin America.[3]

What defines a "mega-city" continues to change as well. In 1950, only eight cities contained populations of five million or more, with New York City becoming the largest with more than 12 million people, followed by London and Tokyo. In 2000, a city the size of New York in 1950 would not be among the world's 10 largest cities. Despite rising populations in existing cities that transform them into mega-cities and the emergence of new large cities in the past century, however, much of the increase in the foreseeable future will occur in smaller cities with populations at or below 750,000 residents. This pattern will occur in countries with developed and developing economies. In fact, despite the alarming growth of the world's mega-cities, they will hold only about 4.1 percent of the total global population in 2015. Although the trends described here have occurred at different times in different regions of the world, with Latin America, Africa, and Asia the most recent to experience explosive urban development, they are an extension of uneven urban growth patterns that reach back into our global past.

As Ian Douglas has pointed out in *The Urban Environment*, the large Sumerian city of Ur, ruled by hereditary monarchs, possessed sanitary

Figure 5.2 The Skyscape of Shanghai, China, the country's largest city with a population of more than twenty million people. *Source*: © Peter Morgan, 2008.

facilities much better than those of the small city-state of Athens with its few thousand citizens 2,000 years later. Subsequently, ancient Romans built major public water works, including the Trojan aqueduct that controlled the flow of potable clean water across the Tiber River into Rome. As a sink for wastewater and sewage, Romans built the Cloaca Maxima, a great sewer, to redirect the natural flow of water out of the city. In the words of historian and economist Joel Mokyr, "The Rome of 100 CE had better paved streets, sewage disposal, water supply, and fire protection than the capitals of civilized Europe in 1800 CE."[4] Yet, despite these technological advances in sanitation, they failed to adequately protect average Roman citizens from outbreaks of waterborne diseases. Providing drinking water for growing urban populations became an acute problem, where providing no more than 7 to 15 quarts of questionable quality to each person a day remained a challenge. Like so many pre-modern and modern cities, huge collections of refuse piled up each day, serving as breeding grounds for infectious microbes carried by legions of flies. And despite the construction of great sewers throughout Rome, most residences above the ground floor and almost all tenements remained disconnected from the sewer system. Public health became a hazard for most urban dwellers in all places, not just the large cities of ancient Europe.[5]

Until industrialization began in Europe in the middle of the eighteenth century, large cities were an anomaly. As world historian David Christian has pointed out, there may have been eight cities in the hub region of Mesopotamia or Egypt with a total of 240,000 inhabitants, with each city holding approximately 30,000 people in 7000 BP. Almost 4,000 years later, in 3200 BP, 16 cities had emerged with a combined population of 500,000 and spread across Eurasia from the eastern Mediterranean to northern India and China. "In 650 BCE, there were still only twenty cities this size, with a combined population still under a million. But by 430 BCE their numbers exceeded fifty, by 100 CE over seventy, and their total populations were, respectively, 2.9 million and 5.2 million."[6] He noted further that a population decline struck urban areas during the first millennium. As a result, "in 1000 CE, there were no more people and no more towns than in 1 CE."[7]

Most of these largest cities were located in regions of the world where agriculture had taken hold in the previous millennia. They included the city of Tenochtitlan, the center of Aztec civilization in Mexico that before the arrival of the Spanish conquerors in 1500 CE contained 500,000 persons.[8] In addition, earlier centers of agriculture located in the Fertile Crescent of Mesopotamia, in Egypt, the Indus Valley, China, and the Mediterranean region of Europe contained some of the world's largest cities, with populations ranging from 100,000 persons in Thebes and Memphis-Cairo in Egypt and Babylon in Mesopotamia in 5360 BP to Rome with 650,000 in 100 CE to Xi'an/Chang'an, China, Baghdad, and Constantinople (Istanbul) with 800,000, 700,000, and 300,000 respectively.[9]

The pharaoh Menes founded Memphis-Cairo about 7100 BP. Its location on the alluvial plain of the Nile River provided water for irrigation and fertile soil for farmland. Protected by large deserts on the east and west and located about 100 miles inland from the Mediterranean Sea, this ancient city with a warm and dry climate was inhabited by 40,000 people within a hundred years after its founding. After a long period of warfare and instability, Memphis reemerged in 5991 BP with a population of 10,000 priests and a total population of about 100,000. During the intervening years, the Babylonian kingdom under Sargon and Naramsin with its capital at Akkad reached the same size as Memphis about 6240 BP. By 6100 BP, its capital had moved to Diakonoff with an estimated population of 100,000. Warfare, chaos, and the rising, falling, and rising once again of Egyptian and Babylonian (Akkadian and

Figure 5.3 Citadel of Qaitbay, located on the site of the Lighthouse of Alexandria, Egypt, one of the Wonders of the Ancient World. Fortified by the Egyptians in the fifteenth century. *Source*: Galen Frysinger.

Assyrian) dynasties led to the emergence of glorious but short-lived cities such as Nineveh in 4668 BP with a population of 120,000 and the renewed splendor of Babylon under the ruler Nebuchadnezzar with a population of 200,000.[10]

Alexander the Great founded the Egyptian city of Alexandria to celebrate his greatness as a conqueror and by 4320 BP it contained 300,000 people. It was matched for a short time by the city of Seleucia in Mesopotamia founded by one of his generals, Seleucus. Further to the east on the Indian subcontinent, the city of Patna thrived as a center of the Gupta dynasty (320–550 CE) that ruled about half of India, with its power concentrated along the Ganges River valley. Patna's circumference was 21 miles and its population reached 500,000 by 450 CE. Rome had grown to 800,000 inhabitants 400 years earlier. After the fall of the Roman Empire in 451 CE, capital cities on the Eurasian landmass gained in population. Constantinople, the center of Byzantium, after the fall of Rome reached around 500,000 in 650 CE. Xi'an Chang'an, China, capital during the Tang dynasty (618–905 CE), peaked at 1,000,000 in 900 CE. Baghdad, population center of Islam, was the world's largest city, with more than a million people by 935 CE.[11]

Despite these ancient urban centers and their important symbiotic relationship with agriculture, the global urban population hovered around 1 to 2 percent of the world's population until the age of industrialization. Population growth and urbanization became spurs to innovations through the commercial trading networks, but pathogens and people became intertwined. Some of the pathogens caused catastrophic epidemics across the Eurasian landmass. With the coming of the age of oceanic exploration in about 1500 CE, disease regimens led to global pandemics. Spreading disease slowed population growth, as many of these pre-modern environmentally unhealthy cities became incubators for lethal pathogens.

The absence of public health lowered agricultural productivity, lowered per capita output, and limited a person's prospects of success. The lack of knowledge about infectious microbes, about the limits of the land's productive capacities, and its needs for nutrients dogged farmers for centuries. Fertile lands that produced bumper crops suddenly declined. The abstract law of "diminishing returns" meant little in a world dominated by farmers accustomed to engaging in a repetitive enterprise. Natural forces beyond one's immediate control, such as the changing climate, the unpredictability of rainfall, the onset of infestations, and erosion, dominated one's ideas about success or failure. In addition, farmers faced feudal social systems and regressive economic incentives that thwarted progress.

At the same time, networks of merchants and craftsmen created competitive markets, established prices, and began to specialize and promote technological innovations. Literacy and accounting methods became valuable skills to be learned, improved upon, and passed on through secular schooling. Although they remained small by present standards and would continue to remain so until the middle of the eighteenth century, evidence of the early beginnings of urbanization were found in the world's earliest cities.

What Does Urban Mean?

According to the noted archeologist V. Gordon Childe, the first cities shared some common features that remain the defining characteristics of all urban space. First, a city possessed a relatively large population within a limited geographical area. Second, this population developed a specialized division of labor and became socially stratified based on the concentration of wealth and political power. Third, with the centralization of power, the leaders of this hierarchical society gained control of the surplus of resources and goods by collecting tribute, levying taxes, and engaging in foreign trade. Trade focused mostly on luxuries for the wealthy landowners, merchants, warrior class, and government officials. Fourth, as power and status became concentrated in the few, it reduced the role of kinship and clan membership and located political representation in entities such as neighborhoods and districts. Fifth, these early cities possessed large public buildings with distinctive architectural designs. Sixth, with the invention of writing for record keeping and contracts, their more learned citizens developed early scientific disciplines, particularly astronomy, mathematics, physics, and engineering. Lastly, art in the form of portraiture celebrated the accomplishments of the merchant class. With these developments, what it meant to be urban was complete.[12]

From the pre-urban world, we have evidence of human populations engaged in collective learning through record keeping. As long ago as 14,300–12,000 BP, hunting and gathering populations were structuring their worlds by the use of notations to signify time and seasonal variations. At the Riparo Tagliente site near Verona in northern Italy, incised notations on a limestone block showed a positional pattern of sets and subsets. To read them arithmetically would reveal little, but to interpret the positional notations as representing the growth patterns of plants and the timing of ritualistic events suggests the improving cognitive and conceptual life of pre-agricultural humans. The ability of humans to organize and sustain increasingly complicated modes of production became prerequisites for the development of agriculture and early settlement.[13]

Early Urbanization and Its Environmental Effects

The beginning of urbanization took place in an environment in which gathering places were identified by Ice Age communities between 30,000 and 12,000 BP for seasonal rituals, for trade and exchange, and for renewal of friendships and the establishment of communal networks.

These locations existed long before semi-permanent and permanent settlements and became the focus of mercantile life. From humble and easily forgettable beginnings, early gathering places became small village enclaves and eventually established towns. In this way, rituals, trade, and exchange represented the early beginnings of urbanization. Among the many early sites, eight urban sites became the centers of civilization. Three dominant civilizations existed in Mesopotamia, Egypt, and the Indus Valley of India. Humans completed their pre-urban phase about 6000 BP with the building of the first cities in the environmentally friendly riverine region of southern Mesopotamia, now modern Iraq.[14]

These early Mesopotamian cities ushered in an urban era of world history and the beginning stages of modern civilization. Neither the development of cities nor the emergence of civilizations would have been possible without significant increases in the production of food. The planting and harvesting of crops along with the domestication of animals created a surplus of food for the first time in human history. Others developed later and independently on the North China plain, in Mayan Central America, in Aztec Mexican cities, in the pre-Incan Andes, and in coastal Nigeria among its Yoruba people. The establishment of these cities marked the end of the extended pre-urban period in human history.

The development of early cities changed the relationship between humans and their environment. Transportation links between the cities and the surrounding areas became significant pathways linking formerly separate entities. In the process, city-building required materials for construction, repair, and replacement. In the earliest cities, mud was soon replaced with weatherproof kiln-fired bricks. Although the kiln technology vastly improved the durability and longevity of urban structures, firing the kiln required vast quantities of wood from near and eventually distant forests. Depleting the forests carried with it destructive ecological practices by destroying complex biological communities. The water retention characteristics of woodlands limited evaporation, protected the soil from erosion, and decreased the velocity of seasonal winds blowing across the land. Creating new urban ecosystems placed unknown stress on rivers, lakes, and streams. Irrigating the land diverted water from its natural ecological pathways, changing the aquatic life of rivers.

The most destructive outcome of early irrigation and one that dogged agriculture since the beginning, however, was the way in which the natural salts from a flowing river desiccate the land, causing long-term declines in agricultural yields. Although less destructive ways exist to treat the land, little of that knowledge was available during this earliest phase of urbanization. The nearly devastating environmental impact of these changes transformed plant and animal communities, changed the quality of the water, and shaped the consumption patterns of humans. As they consumed more resources in nature, they produced more human waste, requiring the development of new infrastructures for delivering water and for removing sewage. Failure to do so systemically resulted in the spread of infection with the attendant increases in morbidity and mortality among a city's most vulnerable populations.

These early cities became organizational centers of economic and political power. With the surrounding areas providing food and people, and natural resources for construction, the evolution of urban life began to unfold. The eventual urbanization of modern society was thousands of years in the making but its roots can be traced back to these early origins. All emerged from their humble beginnings as centers of ceremonial activities and rituals. None of them originally served as military citadels, or as administrative organizations or commercial markets. "[B]eginning as little more than tribal shrines ... these centers were elaborated into complexes of public ceremonial structures ... including assemblages of such architectural items as pyramids, platform mounds, temples, palaces, terraces, staircases, courts, and stelae. Operationally they were instruments for the creation of political, social, economic, and sacred space, at the same time as they were symbols of cosmic, social, and moral order."[15]

Ancient Urbanization

Ancient Mesopotamia (Greek for "between the rivers") represented several thousand years of historical development. Its history began prior to written records, when the small village of al 'Ubaid, dating from the fifth to the fourth millennium and set on a small mound of Euphrates River silt, was established in the river's valley. It extended to the flourishing of Mesopotamia's cities in the second, third, and fourth millennia and its boundaries extended beyond Babylonia proper to include the extensive rain-fed agricultural region of Assyria. The Sumerians of this ancient civilization used the word *uru* to represent a place, regardless of size. It could be a village, a town, or a city. In fact, it was used to identify the southern Sumerian city of Uruk, on the banks of the Euphrates River about 200 miles south of Baghdad, that in 5400 BP had become the largest city in the world and that 800 years later contained within its walls a population of 50,000 living on 1,300 acres. "It consisted of whitewashed mud-brick houses, of a type that can still be found today, with narrow streets running between them. While most were one-story high, wealthier houses often had two stories. In the center, on a ziggurat 12 meters high, stood the White Temple."[16] Its agricultural lands and pastures extended nine miles beyond the city's walls.

The city thrived on the bounty provided by its extensive landholdings. It became the source of the city's water supply, its surplus food, its wood to fuel new manufacturing enterprises, households, bakeries, and pottery-making kilns. Environmentally, this growing city produced rapidly increasing amounts of refuse and waste that became incubators for a host of diseases. With the invention of an urban sanitary infrastructure to carry away both human and animal effluent, some ancient cities escaped the ravages of waterborne and airborne diseases, one of the most common problems associated with urban development. As the largest city in the world until the Roman Republic in the first millennium, Uruk's success was influenced by an almost 1,000-year expansion begun by the building of its temples about 5500 BP. Temples served ritual and administrative functions, organizing colonies and trading posts to the west across Assyria to Asia Minor and southwest into Egypt and across the plateau in Iran.

Similar to many of the large cities that followed, Uruk built warehouses to protect its surpluses of wheat, barley, and rye against attack and famine. It built extensive networks of canals and irrigation works for transport and agriculture. It constructed walls and fortifications for protection from enemies and created a human infrastructure of officials to establish order, maintain domestic peace, and settle disputes. Only growing cities possessed the resources to satisfactorily build, expand, and maintain the physical structures and to deal effectively with the organizational problems posed by internal disputes and aggression from beyond the fortified cities.

The reach of Uruk's urban culture toward its neighbors to acquire metals, stones, timber, and other raw materials resulted in the diffusion of its writing, its cylindrical seals, and its art forms. The city exported its methods of accounting, cataloguing, collecting, and attempts to classify observations from the night sky. Most of what they did emerged from their advanced cuneiform writing, contained in literary texts, proverbs and myths, historical narratives about heroic bravery in battles, sacred hymns, and much more about the daily lives of royalty, elites, and commoners.

Uruk and the other prominent Mesopotamian rival cities of Lagash, Umma, Ur, and Kish led the transition to labor specialization by organizing the urban population of dependent women and children in public weaving workshops. Accelerated specialization in Mesopotamia's ancient cities was identified in the laws of Hammurapi (#274) (3792–3750 BP), which listed a number of specialized occupations organized as special guilds. The list included seal cutters, jewelers, metal smiths, carpenters, house builders, leather workers, reed workers, washers/fullers, felt makers, and doctors.[17]

Economic specialization led to social stratification as temple merchants and administrators engaged in profitable economic activities. As public places for rituals and ceremonies, Mesopotamian temples possessed their own land, animals, and resources to support new and established handicraft industries. Combined with dependent laborers, mostly women and children, these public entities created surplus goods for trade. In the same way, temple ownership of land promoted the first examples of renting to farm tenants in exchange for a share of the surplus crops. In the words of the noted archeologist Michael Hudson, "profit-accumulating enterprise was public long before being privatized."[18] In these Mesopotamian cities, temples became the first public institutions where profit-making and the accounting of costs and revenues represented a public and in some ways a sacred enterprise.

Merchants also initiated foreign trade by establishing trading posts or colonies to purchase precious metals, hardwoods, and stone to manufacture luxury materials to sell to a wealthy foreign clientele. Thousands of years ago, family-centered social arrangements began to be replaced by corporate structures. For example, in Mesopotamia and in Egypt, local communities maintained their own irrigation waterways and dams. Less family oriented and less clannish than previously thought, these early Mesopotamian cities were polyglots, incorporating strangers freely into their societies. One cannot pigeonhole this ancient urban civilization in terms of ethnicity, language, or race. It was truly multicultural in a world that would eventually become stratified and separated by all of the social constructs invented by humans of race, ethnicity, religion, and language.

At many times in Mesopotamian history, invading peoples were absorbed into the larger culture. As a result it became diverse, multilingual, and inclusive. Overrun many times, the Sumerian form of writing adapted many times to the linguistic influences of intruders, from the Elamites of Iran who destroyed the Sumer kingdom of Ur 4,000 years ago to the Akkadians who lived alongside the Sumerians in the heartlands of the two rivers in Akkad around modern Baghdad for thousands of years. By 4300 BP, those who spoke an earlier version of the Assyrian language had forged an empire that extended from the Persian Gulf in present-day Iraq to the headwaters of the Euphrates River in modern Turkey.

Parallel developments occurred among the three ancient urban civilizations of Mesopotamia, Egypt, and the Indus Valley from about 5000 to 4000 BP. About 5000 BP writing unified Upper and Lower Egypt into a single kingdom. Around 4500 BP the era of rapid pyramid-building occurred and dynastic Egypt expanded into Nubia and began foreign trade with Southwest Asia, including the Levant and possibly the Indus civilization. Between 5300 and 5200 BP, writing first appeared in Mesopotamia and major city planning, temple construction, stone sculpture, and cylinder seals became the symbols of wealth, status, and public accomplishment in the Sumerian cities of Uruk, Eridu, Ubaid, and Ur. Colonial expansion along with foreign trade on the Arabian Peninsula and the Persian Gulf reached its zenith with the unification of the Mesopotamian kingdom under the ruler Sargon of Akkad (4334–4279 BP).

Sargon conquered several independent city-states, destroyed their walls and fortifications, and centralized control over these states by making his sons governors. Through centralization, Akkadian rule established formal trading networks throughout the ancient world of Mesopotamia, Egypt, and sub-Saharan Africa as well as Central Asia into the Indus Valley. The Akkadian capital city of Agade served as the central place for this emerging agrarian empire. Newly acquired wealth and power were reflected in the following description of life in Agade from early 4000 BP: "In those days the dwellings of Adage were filled with gold, its bright-shining houses were filled with silver, into its granaries were brought copper, tin, slabs of lapis lazuli, its silos bulged at the sides ... its walls reached skyward like a mountain...."[19]

The sudden collapse of the Akkadian Empire around 4170 BP has been attributed to many factors, including internal corruption, warring factionalism, external aggression, and many others. A body of evidence gathered from chemical changes in the sediment of the adjacent Arabian Sea and Indian Ocean basins, however, suggests that the empire experienced a swift regional change from wet to arid environmental conditions. This sudden shift in the region's climate and weather patterns transformed the lush and productive landscape into a desert. Agricultural productivity, the life-sustaining enterprise of adjacent urban centers, declined and with it the vitality and growth of Akkadian cities.[20]

The Origin of Writing

The origin of writing is a much-debated topic among archeologists and linguists but they agree that it was an essential development in promoting urbanization. Most scholars agree that the urban Sumerians were the first to invent writing about 5200 to 5300 BP. They ruled the lower valleys of the Tigris and Euphrates Rivers in what is now southern Iraq. Egyptian writing may also have begun during the same time period. However, the debate about its independent development or its diffusion from ancient Mesopotamia remains unresolved. Independent developments may have also taken place in China about 3300 BP and by the native people of Mexico about 2600 BP. Most other forms of writing are derivatives of these early independent languages.

The development of writing was so complicated that it took thousands of years before it appeared in its complex form. Very few scribes and members of the elite classes used it to communicate with each other for ritual and accounting purposes. For thousands of years, Sumerians and their ancestors used simple notions on etched clay tablets to record mundane but important transactions such as the amount of grain planted and harvested and the number of domesticated goats, sheep, and cattle born, shorn, and slaughtered. In newly emerging cities, where accounting for agricultural production for sale and distribution to elites and commoners reflected one's power relationship to others, these simply etched markings over long periods of experimentation changed into Sumerian cuneiform. Cuneiform was a writing system that used logograms, similar to pictures, representing words or names and phonetics referring to syllables and letters.

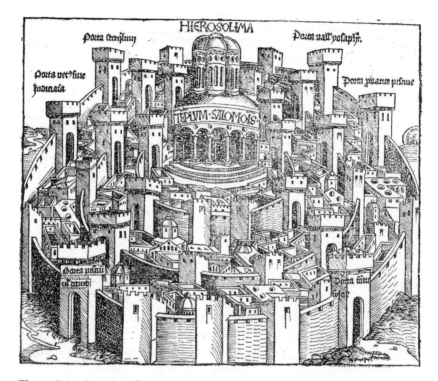

Figure 5.4 An original antique map of Hierosolima (Jerusalem) from the German text "Nuremberg Chronicle" 1493. Hierosolima is represented as a circular walled city with the Temple of Solomon at its center. *Source*: Woodcut by Hartmann Schedel from Wikimedia Commons, the free media repository.

During its developmental stage, the format of cuneiform became institutionalized. Horizontal rows, reading from left to right and from top to bottom, became standardized. With some rare exceptions, almost all modern writing systems throughout the world drew their inspiration from this earliest writing. Among linguists, the origins of Sumerian cuneiform represented the best example of a broader interpretation of writing's origins. Located in the visual arts as pictograms of things in nature and over time, writing developed into more abstract representations of things, names, and finally words in the spoken language. The thousands of clay tablets excavated from Uruk show the earliest development of writing. Over the centuries, their script evolved from clay tablets for record-keeping to visual representations and eventually to a spoken language.

Archeologists believe that writing began as public administrators in the earliest cities sought a mechanism for keeping track of livestock and stores of grain borrowed, bought, and sold. "Record-keeping served as a centralized control and scheduling device long before writing became a vehicle for personal self-expression, literature or abstract philosophy."[21] The clay seals in various sizes and shapes and various forms of notations held temple administrators in the cities accountable for their actions. They served the function of receipts and provided a trail of records that could be reviewed and assessed as time and activity allowed. The keeping of the calendar was another vital urban administrative function maintained by temple administrators.

To make meaning out of the multilingual character of early urban life, Mesopotamian record-keepers invented a common symbolic language of phonetic symbols to translate the foreign names, places, and things. "This is what inspired the transition from word-signs to syllabic cuneiform in multilingual Sumer. Before the end of the fifth millennium writing appeared in all of those places where foreign trade had spread – Syria, Crete, Turkey, southern Iran, the Indus River Valley (now Pakistan), and China. The tradition of writing thus may be attributed to the multiethnic commercial character of archaic cities ... the island that Bronze Age Sumerians called Dilmun (modern Bahrain) when it [was] used to trade with the Indus Valley and the Iranian shore. [The island] served as an entrepot linking Sumer and Babylonia (whose records refer prominently to the 'merchants of Dilmun') to the Indus civilization and the intervening Iranian shore."[22]

Ships of different sizes and shapes able to carry cargo from one region to the next appear as iconographic representations on clay seals, potshards, and models. Cuneiform texts mentioned large quantities of palm-fiber and palm-leaf ropes, suggesting that many ships used them to sew and stitch together planking material. Ancient Mesopotamian texts from Ur list 59,290 wooden pegs for the shipyards of Umma, indicating that, along with stitching, they were used in construction. These texts refer to the use of bitumen mixed with straw, fish oil, and lime to make caulking for watercraft constructed of reeds. Thor Heyerdahl's voyage in the reed boat *Tigris* from Iraq to Karachi and across the Indian Ocean to Djibouti on the Red Sea in 1977–8 confirmed the seafaring capability of these vessels and the trade routes throughout the "liquid continent" of the Red and Arabian Seas and the Indian Ocean.[23]

The Impact of Changing Rivers on Environmental Quality

Despite the fact that many early farming communities began on the hillsides above the bottomlands, from a historical and an environmental perspective many ancient urban settlements were riverine civilizations in fertile zones. The term "Fertile Crescent" can be described in the following geographical terms. It includes the following: land beginning at the head of the Persian Gulf and extending north to the mountainous water sources of the Tigris River; at that point, the Crescent turns west across the Euphrates River through Syria and Palestine. The Sinai Desert interrupts this fertile region before it joins the broad Nile Delta and its narrow valley extending southward into Egypt.

Water played an important role in the daily functioning of these early Mesopotamian cities. For example, the ancient city of Uruk was located about one-third of the way from the head of the Persian Gulf. Baghdad was situated along the western side of the Euphrates River with a wide navigation canal to the north and east of the city. Two harbors located in the north and west of Uruk provided safe harbors for trading vessels. Conceivably, a small canal traversed the city.

Because of meandering over thousands of years, today the river flows about 10 miles to the west of the original city. For example, Uruk's prominence declined and eventually disappeared as the Euphrates River changed course and shifted to the east, leaving the residents of this central city without access to a vital watercourse. Meandering rivers are a natural environmental occurrence, often caused by changing volumes of water in their channels. Their impact on those who live along rivers affected the livelihood and frequently the very existence of a settlement community. While walls and fortifications protected residents from near-constant warfare, natural hazards and environmental disasters placed them at risk without much protection.

Diverting the natural flow of the Euphrates into large urban canals for the multiple purposes of drinking water, irrigation, transportation, flood control, and sanitation posed its own set of potentially adverse environmental effects. Building canals disrupted the natural landscape, altering the habitat for ground animals and fish life in the river. Water temperatures controlled by the river's flow rose in the diversion canals, changing the riverine ecology. During the late summers, the river's flow diminished, creating pools of stagnant water and thereby posing a health hazard for people living along its canals.

Connected river systems attracted people, produce, and livestock and promoted domestic and foreign trade. They had their origins in Ice Age human activities along established trading routes, possessing deep harbors that became commercial centers that stimulated the exchange of goods, ideas, and languages. The centrality of water stabilized cities and insured that agricultural productivity would nourish a growing concentration of urban dwellers. Declining river flow, intermittent droughts, or disruptions of any kind, however, compromised the viability of early urban civilizations.

In the case of ancient Mesopotamia, soil salinization affected the long-range productivity of the land. The sedimentary rocks of the northern mountains leached salt into river water flowing south into Mesopotamia. As water evaporated in the intense heat of the region, salt accumulated in the soil, washed into the water table, or created an impermeable soil surface. The salt in the water table rose to the surface during floods and during irrigation. In effect, it compromised the production of food. Agricultural records for the period note a shift from wheat farming to the more salt-resistant crop of barley. Despite this shift, crop yields plummeted. As with other agrarian civilizations in the world, a destabilization of farming led inexorably to a destruction of the fabric of southern Mesopotamian society.[24]

Urbanization in the Indus Valley

Historically, Indus Valley floodplain civilization is referred to as Harappan, named after Harappa, one of the most important cities of the Valley that flourished from about 4150 to 3750 BP. The emergence of cities in the Indus River Valley benefited from the geological history of the Indian subcontinent, a triangular peninsula between Arabia and Southeast Asia, and accessible once seafarers mastered the navigational winds of the Indian Ocean. The subcontinent's environmental history helped to shape the cultural character of early Indus cities, a history that was largely determined by the interrelationships of the major mountain ranges, including the Hindu Kush, the Karakoram, and the Himalayas that seal off the north, the Tibetan Plateau, and the broad Eurasian plains.

The geological processes that formed the subcontinent's physical geography influenced the region's cultural past and present. The geothermal complexity of the Earth's core over millions of years caused tectonic plates to fracture and combine with other continental fragments, at times creating new continental formations. As noted in Chapter 1, the Indian subcontinent was part of an older super-continent named Gondwana that included Australia and the island of Mauritius.

Within this much longer geological history, the Indus River for thousands of years has served the people along its banks and those who lived on its tributaries as a highway of communication to transport themselves and their goods. Because the river lies between the northern winter rainfall and snowmelt zone of the northwestern mountain range and the summer monsoon, it receives little rain from either. Much like the valleys of the Tigris, Euphrates, and the Nile, the lower Indus Valley is distinctive because a major world river flows perennially through an arid region with almost no rainfall. All of these river valleys served as the centers of early urban civilizations.

The climatic conditions of these ancient regional cultures provided suitable conditions for early agriculture and transportation. In the case of the Indus River, it rises in southwestern Tibet and flows between the two ranges of the inner Himalayas. There, it received ample water from snowmelt and monsoon rains, gaining increased flow from the many tributaries along its course. The Upper Indus was an unstable river subject to the changes in climatic conditions and as it approached its lower reaches, the river changed its course repeatedly because of its shallow channel.

Arid conditions and a flat plain that accepted an increasing amount of silt from the eroding mountains and a meandering river watered the land. While mature forests sprang up along the well-watered plains, the region became suitable for early agriculture as forest cover was replaced with pastures, farmland, and wood lots. The perennial flow of water across the desert plains with annual flooding made possible a bumper harvest of crops. The significance of deforestation, however, on the collapse of Indus cities in the fourth millennium BP cannot be underestimated.

Although the plains and valley of the Indus River in the west receive less rainfall than its major counterpart, the Ganges River in the east, its alluvial fertile soil is better irrigated than the region supported by the Ganges. Conditions in the western Indus basin would have been similar to those found in the Tigris and Euphrates region of Mesopotamia from the seventh millennium BP. During that earlier period, Neolithic farming settlements first appeared on the higher plains of the Indus Valley to avoid the seasonal flooding of the river. With a growing population and the development of technical skills and societal cohesion, farmers confronted the task of cultivating the floodplains.

In environmental terms, the Indus River subjected its principal cities to periodic flooding. They responded by constructing impressive citadels far removed from the lower regions that were most susceptible to the river's raging waters. Built on raised mud-brick platforms and massive walls, these citadels not only protected the elites from floods but also served as ramparts against possible invasion. Walls protected the lower sections of the cities to deter flooding but they seldom prevented the Indus from swamping them.

The long evolutionary period of settlement in the Indus Valley and the rapid spread of the production of millet provided the stability for the early development of urbanization. Although parallels abound regarding the nearly simultaneous development of cities in Mesopotamia and Egypt, their evolution in the Indus Valley was truly indigenous. Even though many of the elements that signify urbanization in Southwest Asia, such as a written language and monumental architecture, seem to be missing from original Indus Valley cities, substantial evidence provides proof of the existence of urbanization.

From 4600 to 4500 BP the emergence of an urban culture took hold in the Indus Valley. New and redesigned cities, beginning with Mohenjo-daro and Harappa, using new ideas about planning and architecture appeared as well as a new and uniform style of manufacturing crafts promoted by leading merchants. Trade between cities and towns in the Indus region flourished, followed by the development of overland trade to the Persian Gulf states and Central Asia. Sea trade with the former developed soon thereafter. As Mohenjo-daro became the principal city for sea trade, Harappa became the gateway to the overland trade connecting the Indus Valley with Central Asia across the Iranian plateau. Yet, archeologists are quick to note that: "Any notion that Indus urbanization is likely to have been the result of dependency on Mesopotamia or [its] institutions is thus based on a misunderstanding of theory."[25] Indus cities, much like its agricultural system, developed distinctly and independently of other ancient civilizations.

The large numbers of stone weights discovered at Harappan sites, particularly its major cities of Mohenjo-daro, Harappa, and Lothal, suggest Indus contributions to the mathematical sciences. "Together they formed a series with ratios of 1, 2, 3, 4, 8, 16, 32, 64, 160, 200, 320, 640, 1600, 3200, 8000, 12,800. One of the most common weights was that with a value of 16, weighing around 13.5 to 13.7gm (0.476oz) … more recently a fresh study of these materials and of further examples from the Lothal excavations [reveal] a new structure, dividing the weights into two series, and producing a striking new symmetrical progression. The first series advances from .05, 0.1, 0.2, 0.5, 1, 2, 5, 10, 20, 50, 100, 200, to 500. The most frequently occurring weight … now appears with a value of 0.5 in this series."[26]

Precise small-scale measurements played an important part in the manufacturing process and provide a window to understanding larger-scale urban planning processes of Harappan systems and other uses of technology. Excavations of urban sites revealed a pattern of different ratios of brick sizes and the relationship of the width of streets to that of houses and their rooms. Such findings suggest the advanced knowledge of Harappan urban planners and the uses of mathematics and geometry in their design of Indus civilization cities.

Organizational symmetry characterized urban form and function. A street plan with a rectangular grid system differed from the patchwork structure of Mesopotamian cities. Harappan cities built sophisticated drainage and sewer systems as well as rubbish chutes to dispose of human and animal waste. Along with their elaborate and secure walls, their great communal warehouses for grains, and the system of weights and measures, Indus Valley cities became well-ordered functioning societies.

As the same time, Harappan cities seemed to inflict a drabness on their residents. "Looking at the way in which the cities were planned, with their rows of mean huts for the workers, grouped close to furnaces for metal production or to pottery kilns, one has the uncomfortable feeling that here were city states in which production was ruthlessly organized but in which the techniques employed were probably not very efficient. One gets, in fact, the feeling that here again the dead hand of the civil servant was in operation, such as was surmised during the declining years of Rome."[27] Polluted air and the material waste from early manufacturing must have been constant reminders to the workers about the poor conditions of everyday life.

Indus cities exhibited no early evidence of writing except for seals and evidence of some town planning along the lines suggested by Indus mathematical and scientific innovation. About 4600–4500 BP the unparalleled growth of large urban enclaves, Mohenjo-daro, Harappa, Lothal, and others, was matched with an equally significant explosion of foreign trade with the cities of Southwest Asia. The archeologist Gordon Childe expressed this parallel development as follows: "India confronts Egypt and Babylonia by the third millennium with a thoroughly individual and independent civilization of her own, technically the peer of the rest. And plainly it is deeply rooted in Indian soil. The Indus civilization represents a very perfect adjustment of human life to a specific environment, that can only have resulted from years of patient effort."[28]

As Childe and others have noted, newly emerging urban centers exhibited evidence of social stratification. The urban phase of Indus civilization provided many examples of differentiation among the Indus population. Prominent architectural buildings including the Great Bath (8 feet deep and 39 by 23 feet on plan) and a warehouse at Mohenjo-daro implied separation among the social classes, prohibiting the bulk of the urban population from the satisfying retreat of the palatial bath and the economic transactions that took place in the city's central warehouse. The evidence of large-scale craft specialization and production in the Indus cities of Mohenjo-daro, Lothal, and Chanhu-daro, built in their own districts, separated from the rest, suggested occupational segregation as well. The production of luxury items for personal adornment such as necklaces, precious metals, and beads signified differences in consumption and access to abundance based on wealth and status.

Mohenjo-daro became the center of Harappan civilization, with its burnt brick wall rising about 43 feet above the floodplain of the Indus River. Further evidence of social stratification within the city was the wide range of housing types for the residents, ranging from single rooms to large homes with many rooms and several courtyards. Big homes built with baked bricks were distinguished from smaller living quarters made with mud bricks. The presence of water for ritual bathing and for sanitation was evident in these early Indus urban centers. The residences of the wealthy, the craftsmen, women, and the poor possessed bathrooms whose wastewater was channeled away from homes by drainage conduits to main drains accessed by manholes below the surface of the streets. In addition to rooms for bathing, many residences possessed privies on the ground or upper floors. Drains for these privies were also connected to main sewers. In some ways this system of wastewater removal remains the same today in modern twenty-first-century cities. The archaeologist Sir Mortimer Wheeler noted that "the high quality of the sanitary arrangements at Mohenjo-daro could well be envied in many parts of the world today."[29]

A method was needed to flush the sewers of Mohenjo-daro. Some archeologists argue that a tributary of the Indus River may have been diverted through the lower section of the city to become a sanitary canal flushing the sewers. Having completed this task, the channel was then redirected to the main downstream Indus River channel. In order to prevent the contamination of the city's drinking water supply that came from private and public wells in the region's porous alluvial soil, a high water table supported by a wet climate and a downstream sewage system seemed essential to maintain the health of the city's population of approximately 35,000.

The decline and eventual disappearance of Harappan culture seems to have come about because of a number of factors, with environmental ones playing a prominent role. In the process of making fired bricks from kilns, the demands for fuel led to extensive deforestation. Overgrazing of the land by domesticated sheep and cattle depleted the soil of its nutrients. More importantly, the geological uplift of the northern shores of the Arabian Sea caused water to flood the Indus Valley plain and inundate the city of Mohenjo-daro. A rising water table accompanied by an increasing salt content in the soil spelled doom for the culture. In the case of Harappa, catastrophic flooding of the Indus Valley accompanied by invasions by nomadic tribes destroyed the highly developed urban culture.[30]

Gradually, these nomads settled down and began to practice agriculture. New settlements, towns, small principalities, and the evolution of new cities, dynasties, and new civilizations were built on former Harappan cities without much knowledge of the region's history.

China's Early Cities

Early Chinese history began in the great bend of the Yellow (Huang He) River and also in the southeast region of the country. From these early enclaves, the Chinese migrated outward in all directions, occupying land and gaining political control over the people they encountered. Chinese civilization began under rulers from a tribe named the Shang whose use of the military chariot gave them a sizable advantage over their neighbors along a substantial area of the Yellow (Huang He) River Valley. Shang China (ca. 3766 to ca. 3122 BP), located on the northern plain, developed the first urban forms. Like so many of the world's early cities, an extensive agricultural base supported Shang-dynasty sites (An-yang, Hui-hsien, Cheng-Zhou, and Luoyang). Village agriculture and workshops for artisans and craftsmen existed in many locales, surrounded by elaborate ceremonial and bureaucratic activities. Since no centralized economy existed for the benefit of the masses, elites living in established complexes benefited from the productive output of the local merchants who produced stone objects, jade, and fine ceramics.

Although agricultural production was dispersed, central authorities controlled the harvest and distribution. Large quantities of stone sickles found at excavation sites suggested a form of centralized control over agricultural workers. At one storage site, no fewer than 3,500 semi-circular stone sickles, both used and unused, were excavated.[31] The existence of storage pits located near ceremonial and administrative sites suggests once again control by central authorities of food harvesting and food granaries.

Additional evidence of centralized Chinese authority during the Shang period is revealed by the construction of ceremonial centers by mobilizing large numbers of workers for building fortifications and cities. Shang China succumbed to another tribal group, the Zhou, sometime between 3122 and 3120 BP. At the capital, Cheng-Zhou (Luoyang), a rectangle of earthen walls each three kilometers long with an original height of 10 meters and a base 20 meters wide surrounded the city. According to one calculation, it would have taken 10,000 workers 330 days a year for 18 years to complete this construction project.[32] For centuries, 3122–2256 BP until the collapse of Zhou China, the dynasty consolidated its power and solidified its heritage by establishing burial rites and producing fine bronze material goods and decorative art manufactured by skilled craftsmen and artisans.

By 2500 BP, scores of Chinese cities existed. Each possessed stamped-earth ramparts and walls for protection from invaders and separate city sections for aristocrats identified by courtyards and larger buildings.

A larger area inhabited by craftsmen and merchants contained commercial streets with shops selling finished goods, including furniture, jewelry, clothing, and food. This larger area also contained taverns, gambling houses, and other forms of recreation. Despite the growth of China's ancient cities, the country remained dominated by farmland and agrarian estates ruled by aristocrats tied to dynastic families.

The introduction of iron manufacturing about 2500 BP, described in greater detail in the next chapter, probably coincided with the rise in productivity. The process of metal casting, namely making iron molds to cast sickle blades and later iron shields, armaments, swords, maces, and other weapons of war, represented inventions in China almost two 2,000 years before they appeared elsewhere. One can assume that the earlier processes of casting bronze and glazing pottery at high temperatures led to casting and forging iron. Tool making was followed by weaponry making.

Regardless of the sequence of events, amply forested and well-watered lands close to these cities crumbled under the demands for wood for fuel and construction. Cooking, heating, metal casting, forging, and pottery making required huge quantities of wood from local and regional woodlands. Desiccation transformed the landscape and forced cities to reach farther away for natural resources. As production costs rose, local economies experienced a decline in the market for their finished goods. For the Zhou dynasty, a period of decline set in with the outbreak of warfare among neighboring states. Fierce conflict among warring factions led to the demise of the small and weak and the rise in prominence of the Ch'in (2256–2206 BP), ultimately resulting in one great empire from which the entire country would be ruled and take its name.

Despite claims that Chinese writing was a pristine invention, scholars hypothesize that it received important influences from West Asian systems of writing. Similarities appear between Chinese script and 22 Phoenician letters. Contact between Southwest Asia and China in the first and second millennia has been verified by the discovery of the bodies of mummies in the western desert of China wearing clothing with Western weaves. These discoveries, along with the acceptance by scholars that the chariot and bronze metallurgy entered China from the West, imply considerable exchanges between the West and the East through an expansion of foreign trade and the rise of urban culture in both regions of the world.

Ancient Mesoamerican Cities

The growth of ancient cities in the presence of agrarian empires accelerated the growth of trading networks, established hierarchical political systems, and promoted the development of languages and collective learning. Their increasingly concentrated populations also placed new demands on the land and water resources, oftentimes leading to degradation and contamination. In Mesoamerica, the growth of cities lagged behind those of Southwest Asia, the Indian subcontinent, and China by at least a thousand years. By 1500 CE, however, the cities of Mexico flourished, with their combined populations amounting to more than 300,000.

The origins of these early Mesoamerican cities may owe their origins to the early cultivation and domestication of maize in Mexico, a process that may have begun as early as 4700 BP. Seven hundred years later, corn began to look much like the food we consume today. As early farmers around the world worked the land, those in Mexico and Central America did the same, selecting the biggest and fruitiest ears for planting and for consumption. This ancient process of artificial selection continues in the modern world, known today as the genetic manipulation of food crops.

The interrelationships between cultivation, domestication, and settlement into villages facilitated the dispersal of corn throughout Mexico and Central America. Along with settlement came other collective enterprises such as weaving, pottery making, and the first semblances of group identity beyond family and kinship. Civic life appeared almost 4,000 years ago among the Olmec of eastern Mexico who built large ceremonial earthen pyramids with monumental sculptures including delicate carvings in jade. Supported by the cultivation of corn harvested four times a year in a warm and wet climate, Olmec culture in Mexico spread in the centuries after 800 CE across Central America to El Salvador.

It disappeared 400 years later but left to its heirs, the Aztec nation of Mexico, an early system of writing based on hieroglyphics and calendars. The deities or nature gods, those large sculptures combining human and predator-animal figures of the Aztec, bear striking resemblances to those of the earlier Olmec. Worth noting was the discovery that Olmec centers contained buried networks of stone drain lines. For example, long U-shaped rectangular blocks of basalt laid end to end and covered with capstones were discovered at the excavation site of San Lorenzo, one of the Olmec's primary settlements. These stone drain lines may have served as aqueducts for carrying drinking water and as conduits for disposing of wastewater.

After the Olmec's disappearance, the Aztec built the first great city in the Americas, named Teotihuacan in Mexico. For three hundred years and like so many of its counterparts in the world of ancient cities, Teotihuacan became a major trading city as well as a religious center, with an array of public buildings and great ceremonial pyramids. It was destroyed around 700 CE, probably by a competing tribe moving south into central Mexico. Until the arrival of the Spanish conquerors in 1500 CE, the history of this entire region is one of competition among regional tribal and ethnic groups for hegemony. Both the Aztec and the Mayan civilizations flourished during an 800-year period following the Olmec.

The Mayan civilization spread across the Yucatan peninsula of Mexico into what is now Guatemala and northern Honduras with neither plows nor the ability to fabricate metal tools to facilitate cultivation in the dense rainforest of Central America. Although the earliest traces of the Mayans date from 2400 BP, their civilization flourished between 1400 and 1100 CE, a period in which they built cities with great ceremonial centers that contained combinations of temples to their gods, pyramids, tombs, and athlete courts for ritual competition. Their stone hieroglyphic writing provided a window into Mayan culture and its people.

During their golden age, Mayans constructed monuments every 20 years to mark the passage of time. They possessed a concept of time calculated in terms of hundreds of thousands of years. This concept of time and the invention of a calendar that used astronomical observations and calculations may represent its civilization's most lasting contributions to human history.[33] Along with constructing Chichen Itza, a city that contained at its zenith approximately 40,000 people with a rural population 10 times as large, Mayan civilization left a lasting imprint on world history.

At its zenith in terms of size, power, and cultural dominance, between 8 and 10 million people lived within its sphere of influence. The matter of how the Mayans supported and sustained a population of this size, not achieved again until the twentieth century, remains a subject of scholarly inquiry. This interest intensified with the discovery that by 1000 CE the Mayans had abandoned most of their largest cities, leading to the collapse of their civilization. Without evidence of an enemy invasion, internal factors of environmental degradation provide an explanation. The Mayans practiced slash-and-burn agriculture in their amply watered tropical forest regions of Central America. In a tropical climate, this agricultural practice turned burned vegetation and the slashed biomass into fuel for abundant crop yields of corn, squash, beans, and gourds. High crop yields will support a growing population with an even larger population requiring even larger future yields. This feedback loop demanded more slashing and burning of the forest canopy.

Degradation of the land caused by aggressive agriculture placed excessive demands on the region's ecosystem. A system of agriculture that was suitable to the population demands of approximately 25 people for each square kilometer could not sustain a population of about 250 people living on the same area.[34] Deforestation of the tropical landscape for agriculture as well as the construction of towns, cities, and elaborate ceremonial and governmental structures created the conditions for an eventual collapse. Evidence of deforestation, however, extended backward in time from 4000 BP to 3000 BP, even though the most significant transition from woodlands to open vegetation occurred between 1 CE and 1000 CE. So, decline was not immediate and it was not caused by a single condition.

Declining crop yields due to years of agricultural exploitation caused food shortages, eventual famine, political unrest, and abandonment of the land. All were important factors in the decline of Mayan civilization. Exacerbating these human impacts, a change in the climate may have also contributed to the decline. Coinciding with human exploitation, a drying of the normally wet tropical climate placed added pressure on agricultural production. Within a few centuries after the changing climate, archeologists believe that the region's population may have fallen by as much as 80 percent.

Early European Cities

Ancient European towns and cities flourished throughout the Mediterranean basin as Greeks and then Romans built ships for trade and conquest. Limited urbanization along the coastal zones became the core of ancient Europe. For the Romans, the rivers in central Europe represented the outer reach of their influence to the east. In the west, England and France contained few urban settlements. At the zenith of Roman hegemony in western Europe, its towns seldom contained more than 200 people. Since slaves mostly maintained the empire's fragile infrastructure of urban workshops and rural landed estates, when either civil unrest or invasion threatened Rome these weakened urban settlements collapsed. "The distinctly urban way of life with its special contributions to civilization had not survived."[35]

With the collapse of Rome, a more stable structure supported early European cities from 600 to 800 CE. The heavy metal plow, the horseshoe, and the harness as well as crop rotations improved the ability of farms to support new towns and cities. Farm surplus, always an unpredictable and at times unknowable variable on Roman estates, encouraged more intensive farming. With rising farm productivity, populations grew and more non-farm producers moved to growing cities. As the migration of farmers from beyond the older Roman boundaries continued, new farming technologies replaced conventional methods. With additional surpluses, the incentives to move to towns and cities to produce for both internal markets and for regions beyond the locale accelerated. In this emerging medieval urban complex, warriors in castles needed new armor and war-horses. The nobility desired luxury fabrics and spices while peasant farmers for the first time could afford cloth, salt, pots, and tools.[36]

Although early European society would remain dominated by agrarian interests, its towns and cities became dynamic centers of pre-industrial manufacturing and economic and social innovations. By 1100 CE, thousands of these new urban configurations appeared on the European landscape. Many of them developed a political and administrative independence. Yet, for most of the pre-industrial centuries, they depended on agrarian feudal landholders to supply burgeoning towns and cities with farm surpluses and the natural resources to construct and maintain these new entities.

The demand for resources would grow as new towns and cities were established and older ones expanded. Woodlands provided construction materials, and fuel to heat residential and commercial lodgings and to power the workshops of smiths, bakers, tanners, and others. Although deforestation was a byproduct of expanding agricultural activity to produce surpluses for towns and cities, the construction and fuel demands of these early cities taxed the surrounding resource base even further. With deforestation came the unintended effects of soil erosion, silting of the adjacent rivers and streams, and more volatile changes in microclimates.

As described by the economic historians Hohenberg and Lees in *The Making of Urban Europe, 1000–1950*, the early history of the English provincial town of Leicester provides a profile of urban development before the age of manufacturing and industrialization in the eighteenth century. As described by them, the Romans built a fort there about 100 CE at the crossroads of one of their major roads, Fosse Way, and the Soar River. In the centuries that followed, the location became the site of a Roman administrative center with fortified walls, a public forum, and baths. Although little is known about the town's history for the 200 years from 400 to 600 CE, a Catholic bishop established Leicester as the center of his diocese in the seventh century. With the establishment of the Anglo-Saxon kingdom of England, it divided the country into several counties, with Leicester becoming the capital of Leicestershire, a role that it retains today.

After the Norman Conquest of England in 1066 CE, King William (of Normandy) ordered the construction of a castle in Leicester. In 20 years, the town had grown to include 378 houses, six churches and a population of about 2,000 persons. The growth of the city continued without interruption through 1200 CE. By then, the town was divided into nine parishes (districts), with some beautiful building designs found in its local churches and public buildings. As a city dominated by commerce and serving as a marketplace for Leicestershire, it produced bread and beer for local consumption and its merchants remained major exporters of wool. Tanning leather and manufacturing leather goods, from saddles and harnesses to boots and apparel, continued to flourish into 1400 CE. By the 1500s the town contained a number of skilled craftsmen, including shoemakers, tailors, weavers, and bakers. More than half of its residents worked in manufacturing while another 25 percent were farmers producing food and milk.

An organization known as the Guild Merchant controlled the town's commerce, erecting barriers for entry into the marketplace and transacting much of the town's financial business. Leicester's political power rested in the hands of the Earl of Leicester, appointed by the king of England, and elders who handled the town's civic matters. By 1500 CE elite members of the town's wealthy families consolidated their hold on the government. And after the earl's estate became part of the king's landholdings, a closed corporation controlled the town's political and economic affairs as well as the selection of members to Parliament. Throughout much of Europe, governance remained the privilege of the few rather than an entitlement of the many.

The case of Leicester's urban development followed a familiar pattern of town development in medieval European history. Although fewer than 5,000 residents inhabited most towns, some, such as Florence and Milan in Italy, contained more than 50,000 inhabitants living in multi-story buildings or in neighborhoods beyond the city's walls. By 1300 CE, Milan controlled 50 additional towns and 150 villages, while Leicester's political boundaries were fixed despite the reach of its commercial enterprises into the surrounding region.

So, it would be folly to attempt a simple categorization of European medieval towns and cities. Although Florence and Milan were distinctly different from Leicester, the bustling port cities of the Mediterranean and the imperial capital at Constantinople, which grew from fewer than 100,000 residents in 1453 CE to as many as 800,000 in 1600 CE, shared qualities unknown in other settings.[37]

The expansion of many European medieval cities was the outgrowth of "prodigious feats of civil engineering."[38] Whether their deepwater ports, their qualities as hubs or magnets for surrounding market economies, or as agrarian centers, their growth and development required a commitment of staggering proportions. Surrounding woodlands and quarries provided the construction materials for continuous expansion. Dredging watercourses and harbors altered the landscape with piles of silt removed from the waterways. That much of the silting resulted from deforestation that loosened the soil probably escaped the thinking of most medieval observers. The dredging ended when either the task became overwhelming or the site lost its economic viability. "The decline of Bruges (in Belgium) in the late Middle Ages as a major trading center has been attributed to silting in the channel leading to the sea, but it is far more reasonable to say that regional shifts in trade, which reduced the city's role, discouraged the burghers from regular dredging of the waterways."[39]

For hundreds of years, the majority of European towns were local market centers with fewer than 2,000 inhabitants. By 1300 CE, between 75 and 95 percent of the towns in France, Germany, and Switzerland fell into this category.[40] Few survived the economic depressions of

Figure 5.5 The Building of the Avenue de l'Opéra in Paris, France was completed in 1878 as part of Napoleon III's revitalization of the city. It remains a focal point of the city's business, tourist and interurban transportation system. *Source*: Photo by Charles Marville, 1816–1879. © Musee de la Ville de Paris, Musee Carnavalet, Paris, France. The Bridgeman Art Library.

the sixteenth century but others continued to attract residents by providing larger economic markets for skilled weavers, tailors, and bakers. Others became county seats for the region's administrative, financial, and military activities. Some Italian communes such as Pistoia in 1300 CE ruled over an area of some 300 square miles, collecting taxes, requiring military service, maintaining public roads, and controlling hospitals and churches. Once it came under the control of the commune of Florence in 1350 CE, it lost these powers to its new protector.[41]

The northern cities of Italy – Florence, Venice, and Genoa – and the Low Countries cities of Flanders and Bruges contained city populations of 10,000 or more. London with its 50,000 residents in 1300 CE and Paris with about 80,000 in 1340 CE proved to be the exceptions. The rest of Europe remained an environment of small villages and towns. With the growth of market cities that linked central places and/or became pieces of larger commercial and political networks, they began to extract more and more resources from greater distances.

According to Hohenberg and Lees, this expansion may have begun when medieval gardeners began carrying topsoil to village gardens from surrounding areas. As these communities grew in size, grain, vegetables, and livestock for poultry and meat were produced further away from the local environment. Later urban growth required fossil fuel and mineral fertilizers dug from the ground to accelerate future developments. With increasing dependence on non-renewable natural resources, the ecological footprint of cities became broader and deeper.[42]

A perception that urban growth was a gradual long-term process is misleading because growth was sporadic and punctuated with periods of expansion, collapse, and gradual change. Only in Great Britain, including Ireland, was urban expansion truly a long-term, gradually accelerating process. For example, London grew from 40,000 to 575,000 from 1500 to 1700 CE. For the century from 1550 to 1650 CE, the pattern of growth among Europe's cities accelerated only to slow down to a trickle by 1750 CE. A gradual recovery of urban growth in Europe occurred later in the 1700s. After that the growth of the total population stagnated. Mediterranean population expanded only to collapse in the 1600s while in northern Europe major gains in urbanization occurred by 1650 CE.[43]

When urban Europe is examined in the aggregate, developments that began in the 1600s changed the character of the continent and its relationship with a global network of trading partners. In the period from 1600 to 1750 CE, 38 cities accounted for 80 percent of the urban growth on the continent and 30 of these were either capital, port, or national trading cities. Almost all remained the great capital cities of the twentieth century.[44]

Despite the growth of these large cities, maintaining the momentum of urbanization from 1750 to 1850 CE was triggered by the growth in the number of small cities with populations of 10,000. Their numbers tripled during that 100-year period, some as the result of the creation of new factory towns but many others by the impact of economic and population trends already evident in Europe. In 1750, there were 138 such cities and a century later 551. Forty years later in 1890, as a result of the trends noted earlier, the number of cities with populations ranging from 10,000 to 19,000 almost doubled to 1,024 cities.[45] In Britain,

50 percent of the population lived in cities by 1850; Germany reached this percentage in 1900, the United States in 1920 and Japan in 1930.[46]

Although the age of manufacturing began before the emergence of cities, much of it took place in relatively small locations with few people. Driven by technological innovation, small cities and rural communities were located near the source of mineral wealth. They attracted more people for productive work, causing a demographic expansion. In many rural towns and villages, manufacturing spread beyond the household and the workshop to encompass a larger geographical area. Beginning as early as 1200 CE, new cities in Europe became the springboard for early manufacturing.[47]

CHAPTER SIX

MINING, MAKING, AND MANUFACTURING

Introduction

The age of manufacturing began many millennia ago when humans began to fashion tools, ornaments, and weapons from stones. Shaping stones by striking them with larger stones, rubbing them to create a sharp edge, and eventually drilling holes to make a shaft for a wooden handle represent the earliest attempts to use materials from nature for protection, work, and ornamentation. The age of stone, or the Paleolithic Age, came to an end as humans began gradually to make the transition from stone to metal tools, ornaments, utensils, and weapons. The use of wood for all of these purposes transcended the ages of stone and metals as humans used wood for all of these purposes throughout early human history.

This chapter and the one that follows examine mining, manufacturing, and industry as continuities rather than as historical departures. Mining and manufacturing spanned thousands of years and their impact on the natural world pales when compared to the environmental changes of the industrial age of the past three centuries. Although considerably smaller in scale and population than industrial ones, manufacturing societies managed to produce a considerable number of durable goods for useful purposes, make them available to a larger population, and develop systematic local and regional trading networks. In the process, they unleashed pollutants that had local, regional, and hemispheric environmental effects. In this way, manufacturing became a precursor to industrialization. Unlike the gradual rise of early manufacturing, however, the transition to the industrial age began more abruptly with the widespread use of coal in sixteenth-century England. It became the energy source that revived stagnant industry caused by a fuel crisis during the long history of manufacturing.

Agricultural productivity became a prerequisite for the expansion of manufacturing because it released members of a growing population to engage in human activities other than hunting and gathering food. With agriculture came early settlements, towns, and eventually cities where a growing occupational and social differentiation among people became apparent. Agriculture and metallurgy also fostered the beginnings of urbanization. A growing differentiation among the population with a range of social and occupational classes became one of the distinguishing features of ancient cities. Artisans, metal smiths, skilled, semi-skilled, and common laborers represented members of an emerging economy in manufactured goods.

The synergy between early agriculture and metallurgy fostered economic and social development. Cooking in a hearth became a central household activity and skills acquired from tending hearth fires and cooking at different temperatures were then adapted to the practice of metallurgy. Prior to advances in metal technology, households produced manufactured tools, ornaments, and utensils for local consumption. Making ceramics, initially by hand and eventually using the pottery wheel, remained essentially a kinship, non-specialized form of manufacturing.

It coincided with an early agricultural regime based on hoe cultivation using human muscle power, not the energy of animals. This initial stage of agricultural development in which domesticated animals were consumed primarily for their meat was replaced by a secondary products revolution. This transformation took place 6000 BP with the invention of the plow and the cart as well as the breeding of domesticated cattle, sheep, and goats known for their meat but also for their secondary products of milk and wool.

Each of these inventions, the pottery wheel, harnesses for oxen and horses, animal breeding, and the secondary products revolution, changed the relationship of humans with animals and with nature. More cultivated land for food and fodder required clearing the land of forests, scrub, and brush, exposing ground cover to direct sunlight. Eliminating the canopy, the land's protective cover, changed microclimatic conditions. Natural wind breaks disappeared, as exposure to wind chill during the coldest of winters made local environments inhospitable to humans, animals, and the wellbeing of the microorganisms that renewed the soil's fertility. At the same time, the increasing use of animals for food production and as alternatives to human labor

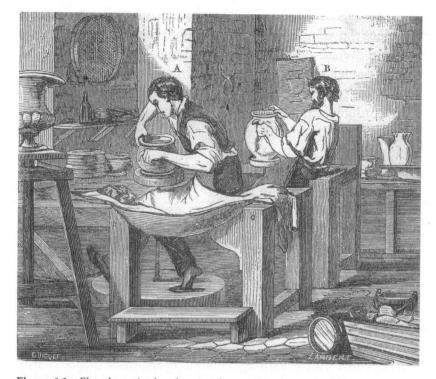

Figure 6.1 *Ebauchage*. A: *ebauchage sur le tour*; B: *tournassage* (turning on a potter's wheel). *Source*: From the 1864 book *Dictionnaire de chimie industriel*, vol. 3, p. 306.

fertilized the land with their manure and urine. An organic exchange of energy between humans, animals, and plants would accelerate as farming and manufacturing became more intertwined.

The invention of the heavy plow by the Romans in 100 CE and the iron plow about 1000 CE removed many of the environmental constraints of localized, family-based, hoe-driven agriculture. The manufacture and increased use of metal tools, including the ax, transformed local environments. Advances in agricultural productivity and in the production of metals encouraged settlement and the emergence of ancient cities. In turn, they led to the rise of complex social organizations and created opportunities for regional exchange networks. With the expansion of these networks, information began to flow across a much wider territory. Flow, exchange, and information processing expanded the knowledge and communication apparatus of traders, producers, and consumers. Social interaction among previously isolated bands of people intensified while control over their environments expanded. Civilization, as we have come to know it, began its slow uneven development.

The Age of Copper and Bronze

When exposed to the elements, copper is easily identified by its bright hues of blue and green. Some of it was found in glacial moraines and along the boundaries of tectonic fractures, signifying the important relationship between a natural world evolving many millions of years before humans and their inventive capacity to transform natural materials into tools, weapons, utensils, and ornaments. Coming to know materials in nature, to work, mold, and manipulate them, expanded knowledge. Metallurgy and manufacturing became unified human activities.

Copper dominated the world's production of metals from 9000 BP to about 3500 BP and remained so in the western hemisphere before the arrival of Columbus. Found in many worldwide locations, copper is a thousand times more common than gold.[1] In geologic time, clashing tectonic plates created the physical environment in metamorphic rocks that became rich in iron oxides. By chemical processes, copper filled crevices in the surface, creating identifiable veins, nodules, and leaves. Over millennia, geological uplifts created large valleys that filled with alluvial sediments to cover large mineral deposits in Southwest and West Asia. Copper tools and weapons replaced Stone Age implements and in turn were replaced by Iron Age inventions.

Although the development and coordination of copper manufacturing was probably quite slow, once installed, it became a complicated endeavor requiring the mastery of several processes and the coordination of many people. Its first uses appear in the millennium from 10,000 to 9000 BP when heating and hammering shaped copper into simple tools and weapons. By 7000 BP, humans, producing temperatures of 1950 °F, cast copper from surface deposits of ore. Archeologists have recovered copper rings, chisels, axes, knives, and spears from early Mesopotamian villages, predynastic Egypt, the Mohenjo-daro society of the Indus Valley, and ancient China. Once humans exhausted surface deposits of copper, deep mining became a labor-intensive activity, some of it carried out by slaves.

Although humans had excavated land for burials and dug deep settlements to protect themselves from the weather, deep mining became a new human experience. We know very little about these early miners, the effects of work on their health, and the technology of mining. We know, however, that mining began as early as 5000 BP in eastern Serbia where miners worked at depths of almost 100 feet digging with wooden levers, stone hammers, and deer antlers. At about the same time, Great Lakes Indians worked at depths of 20 feet extracting at least 3,000 tons and perhaps as much as 50,000 tons of copper.[2] Other ancient copper mining centers have been excavated in the Sinai Peninsula, North Africa, Syria, Iran, and Central Asia. Later discoveries appeared in Italy and Iberia.

In India, a number of prosperous towns flourished near the copper mines of Rajasthan from 2400 BP to 400 CE. Copper coins as well as the residue of mining operations littered the land. Rock, ore, and tailing dumps as well as the remains of smelting furnaces, crucibles, and washing tanks are reminders of old working mines spread out over a two-mile area. Based on the archeological record, the mining and metallurgy of lead, zinc, silver, and gold began there in 2200 BP. Although gaps exist in the chronological record about the mining and working of these metals, as late as 1600 CE the price of Indian copper rose more rapidly than the price of silver as the demand accelerated for cannons, household utensils, and drinking vessels.[3]

In Europe, Roman mining and metallurgy benefited from a long history of discovery, innovation, and transition from copper, to bronze, and finally to iron. Copper use began at least 7000 BP with significant copper deposits located on the island of Cyprus. In the Mediterranean region and in Southwest Asia, copper was easily identifiable by its reddish color or as weathered blue-green. As a soft metal, pre-modern people almost everywhere shaped it easily. In Mesopotamia, the Indus Valley, and especially in the city-states of Sumeria and Happara, fashioning and using metal objects were commonplace. Their discovery and use, however, were not limited to any particular region. Copper use extended to England, western France, Italy, Spain, and to central Europe, before these geographical areas possessed their current names.

Sumerians transported copper from Armenia in the south along the Euphrates River to its major cities in northern Mesopotamia. The word "urudu" in Sumerian means copper, as does the river, Euphrates, so one could assume that the term "copper river" possessed economic meaning for entrepreneurial Sumerians. By 5500 BP, basic copper technology had arrived in Egypt and copper tools for working stone began within a few hundred years as tombs and monuments became signatures of Egyptian society. As Egyptians began the era of constructing great buildings and pyramids, the ancient world's largest copper-mining operation began in the Sinai Peninsula in 5150 BP and continued for almost 2,000 years until the reign of Ramses III in 3186 BP. Its operation met the growing demand for copper tools, including axes, knives, and other sharp implements.

The clearing of forests for farming, for fuel, and for settlement opened land for livestock. The slow replacement of pigs with cattle signaled the replacement of closed woodlands with more open landscapes. The addition of the copper ax to the toolkit of early farmers and loggers may have accelerated the process of clearing forests as well as the expansion of a community's mining capacity. By extracting metal from minerals to manufacture tools, early cities were able to expand geographically and demographically, to place more land under their control, and to extract more from its land and soils. Five thousand years later, they mass-produced copper from sulfide ores.

Pack animals became an important addition to the development of mining and copper exchange networks. Ceramic models of donkeys about 3400 BP provide evidence of the early mining and transportation of copper. The spread of copper technology throughout the Mediterranean region followed, with advances in the smelting and fabrication of household utensils, tools, ornaments, and weapons. The archeological sites of ancient civilizations in Southwest Asia, Cyprus, Anatolia (a part of modern Turkey), and Crete provide many examples of one civilization capitalizing on the advances of an earlier one. Advances in smelting never compensated for copper's perceived weakness, however. Instruments and weapons were never strong enough to hold a sharp edge. As a "soft" metal, hammering it caused it to become brittle and to break.

So, annealing commenced, namely the heating and slow cooling of metal to strengthen it and prevent brittleness. Casting melted copper followed, and finally metal smiths invented cast bronze by combining copper and tin. The reasons for these transitions and the processes that led to their development and widespread use remain subjects for discussion and debate. Some metallurgists suggest that mixing tin with copper to create bronze occurred accidentally. Despite the absence of any evidence that describes exactly how Sumerian metallurgists originally fabricated this new alloy, archeologists believe that metallurgists "stumbled" onto this discovery. Further developments of strengthening and hardening this new alloy resulted from conscious experimentation, however. If human knowledge initially advanced by accident, it progressed rapidly by experimentation.[4] This experimentation with readily available copper and scarce deposits of tin provided a solution that ended the age of copper and ushered in the age of bronze. As an alloy, optimally composed of 90 percent copper and 10 percent tin, experiments to achieve the strongest bronze implements took centuries to achieve.

Elaborate tin mines have been located in the central Taurus mountain range of south-central Turkey dating from 5750 BP. The Kestel-Goltepe mine yielded large quantities of tin and a processing complex for smelting tin existed at the large walled settlement of Goltepe from 5290 to 3840 BP. There, archeologists unearthed thousands of tin-processing stone tools, numerous crucibles, and kilos of powdered tin ore. The mine contained many steep shafts; some so narrow as to suggest that only children could work there. Skeletal remains of children found during recent excavations confirmed this conclusion.

The size of the mining and smelting operation can be calculated by the amount of waste created and dumped on the land. According to archeologists and geologists, one large heap alone contained an estimated 600,000 tons of slag. The debris left at the Kestel-Golpete facility is another vivid example of mining/metallurgy production in the ancient world and their tin slag heaps are reminders of a blighted landscape. Since its mines produced no copper, its tin was mostly exported to other cities for making bronze. Despite these fascinating discoveries, little is known about the actual amounts of tin excavated over time, the identity of the miners, and the producers and consumers of tin/bronze goods. Many archeologists regard these discoveries, however, as an important resource for understanding the technological and cultural development of the ancient bronze civilizations of Southwest Asia.

In the excavated tombs of royalty and nobility in ancient Sumer (4650–4500 BP), riveting and soldering with tin was used in the bronze casting of life-size human statues. Bronze weapons became the weapons of choice among warring factions and conquerors. Around 4250 BP Sargon I united the Sumerian city-states into his kingdom of Akkad and later invaded Anatolia to gain access to its tin mines. Allegedly, one of his caravans alone carried 12 tons of tin, enough to manufacture 125 tons of bronze weapons, bronze-sheathed armor, spear points, and swords to equip a substantial army. Despite his power, the kingdom collapsed 200 years later in civil war. The subsequent emergence of Babylon under Hammurapi lasted until 3900 BP and was replaced by the Assyrian Empire.

Assyrians dominated the region for many years to come. With its political and economic capital in Assur, it shipped tin to its "manufac-turing city" of Kultepe, a city near sources of fuel and copper and equipped with skilled smelters and metal smiths. There, they fabricated bronze weapons, tools, ornaments, and statuary. Although we lack spe-cific information about the volume of work performed there, the city may have been one of the urban world's first "smokestack" cities whose polluted air contaminated the surrounding countryside and posed a constant health hazard to its inhabitants.

Weaponry and trading vessels became the symbols of power, first in the eastern Mediterranean and later throughout the entire region. Bronze weapons spread by Mycenaean and Phoenician merchant fleets reached Carthage in northwest Africa and then Spain in 3100 BP.

Carthage eventually challenged Rome for dominance over the Mediterranean region. In order to maintain control over one's territo-rial possessions and to expand into new areas challenging the current regime, one's smelting operations had to be constantly replenishing existing supplies of bronze weapons, protective armaments, tools, and implements. Not only would it tax supplies of copper and diminishing amounts of tin, mining operations would scar the earth and cover it with debris and contaminants. The smelting activities that followed sent contaminating metal residues into the adjacent waterways and oxidized pollutants into the atmosphere.

Ancient mining altered the local and regional environments. In addi-tion to undermining the stability of the ground, sometimes causing slippage, subsidence, and collapse, mines produced enormous amounts of waste that cluttered the landscape. Mining, however, represented only the first laborious task in a complicated set of manufacturing activities. Charcoal production, then and now, required ample wood-lands and numerous facilities for fuel production as well as smelting and casting. In general, the production of copper and later the alloy of bronze from copper and tin production consumed enormous amounts of timber and caused widespread deforestation.

The production of charcoal required the cutting of extensive woodlands and the burning of approximately seven tons of moist or wet wood to produce about one ton of charcoal because so much energy was consumed in the production process of simply drying the wood. For dry wood the ratio became 5 to 1. Charcoal is produced by breaking down the chemical structure of wood under high temperatures in an earth-mound kiln. Once the process is complete, workers remove the charcoal as the kiln gradually cools down. Typically, around two-thirds of the energy is lost in the process.

Each transition from one production process to another and from one metal to the next represented a synthesis of metallurgical learning. Advances in mining technology coupled with making charcoal to produce metals required building molds for casting, erecting furnaces, creating sufficiently high temperatures, designing ingots for cooling, and transporting bulk copper. It also reflected an advancing level of social complexity because of the need to coordinate the production process. The entire process required levels of cooperation and coordination among many individuals and among various social and occupation groups.

The Effects of Ancient Mining on Human Health and the Environment

During the Bronze Age (4300–2700 BP), land clearing for agriculture, as well as for the mining, smelting, and metal industries, overwhelmed the forests of Southwest Asia. As the primary fuel source, timber served also as an important construction material for ships and for growing towns and cities. The ingredients for making plaster, cement, and bricks used for the construction of private dwellings and public buildings came from the ash residue of burned wood. These demands placed further pressure on the region's woodlands. And once entrepreneurs and those in their employ exhausted the sources of supply near the construction sites and smelting furnaces, they replaced the supply by importing logs from surrounding mountain areas. A cut forest seldom regenerated quickly in this mostly dry and hot Southwest Asian climate.

As we know, deforestation exposed the topsoil, creating extreme climate conditions. The noonday sun baked it and infrequent but devastating torrential rains washed away unprotected soil. Excessive heat and precipitation promoted erosion.

Evidence of deforestation and soil erosion in Southwest Asia appeared as early as 3200 BP and began to spread to the rest of the Mediterranean region. For example, the Laurion silver mines produced 3,500 tons of silver and 1.4 million tons of lead during a 300-year span for classical Athens, beginning in 2478 BP. To achieve this output, smelters burned one million tons of charcoal and depleted 2.5 million acres of forests. Deforestation and soil erosion were so extensive that Plato, when writing about this calamity, said:

> a mere relic of the original country ... What remains is like the skeleton of a body emaciated by disease. All the rich soil has melted away, leaving a country skin and bone. Originally the mountains of Attica were heavily forested. Fine trees produced timber suitable for roofing the largest buildings: the roofs hewn from this timber are still in existence.[5]

A timber crisis spanned time and place in the ancient world of copper and bronze smelting. Once available supplies of wood were used up, ancient Athenians and Spartans exploited northern forests in Macedonia and overseas supplies from southern Turkey, Lebanon, and Italy. By 2400 BP, Athens removed its smelting operations from the mines at Laurion to the coast to avoid excessive transportation costs. Barges filled with charcoal fed the furnaces. Elevated lead content in the slag heaps at these newer facilities, however, suggested that ore was smelted with less fuel, a sign of charcoal's scarcity and expense. The release of toxic lead oxide into the atmosphere will be discussed in greater detail later in this chapter but suffice it to say that it negatively affected the health and wellbeing of smelters and smiths.

A fuel crisis created by widespread deforestation curtailed and eventually terminated smelting operations throughout the ancient world. The Romans stopped smelting ore on the "the smoky island" of Aethaleia in 2100 BP because of declining woodlands. In modern history, the environmental damage done to this same island, now named Elba, is etched in its landscape. Magnificent pine and deciduous forests disappeared throughout the region. The copper industry on the island of Cyprus consumed 200 million trees in order to produce about

200,000 tons of copper. This timber total translated into woodland 16 times the area of the island.[6]

Five thousand years ago, large cedar forests in Lebanon disappeared as the logger's ax provided timber for shipbuilding and for residential and public construction. The fuel consumption of many large and small smelting furnaces caused much of the deforestation of Cyprus, Greece, and the Mediterranean seaboard. The systematic removal of its vegetative cover exposed the land to wind, rain, and unfiltered sunlight. Soil erosion and runoffs into the streams, rivers, and coastal ports caused silting and clogging of natural waterways. In areas with higher population concentrations, residents built levees and diversion canals to control the flow of water.

Once available mature trees disappeared as a fuel source, the production process stopped, despite the presence of ore. Smelting, a process by which ore is melted at high temperatures to produce the desired metal, produced slag. As a fused glassy waste material, slag is created when metal separates from its ore. Dumped in piles around production facilities, slag leached toxic chemicals into the soil, contaminating the land and water.

Copper slag represented a health hazard for miners and local residents. It contained a high chloride content and trace amounts of carbonates, sulfates, arsenic, cadmium, and lead. All are toxic and some are carcinogenic, posing threats to pregnant women, fetuses, newborns, and all persons subjected to exposure over time. In addition, arsenic vaporizes during the smelting process, contaminating the air supply for workers and the inhabitants of nearby villages and cities. Although ancient settlements remained small when compared to modern size and density patterns, toxic fumes, smoke, and fire compromised the health of ancients who lived in the vicinity of copper smelting workshops. If we can assume that much metal processing took place in both small and large ancient cities, then its pollution exposed thousands of people to toxic residues.

Despite environmental and human health costs, the production of copper, bronze, and other alloys expanded throughout the ancient world either through diffusion or through independent invention. A number of manufactured weapons, utensils, tools, and ornaments have been found in ancient civilizations from several regions of the world, including the river valleys of the Tigris–Euphrates, the Nile, the Indus, and the Yellow (Huang He)–Yangtze (Chang Jiang). They run the gamut of weaponry, including lances and arrowheads, knives, and daggers of many shapes and sizes, including those with turned-up curved blades and swords. Tools included saws with teeth much like modern ones, hooks of all shapes and sizes, sickles, chisels, awls, drills, needles, scale-pans, and spatulas. Archeologists have uncovered bronzed polished mirrors, the most distinctive of objects, along with many decorative ornaments of copper, bronze, silver, and gold. Thousands of objects, emblematic of the copper and bronze ages, suggested much about evolving and socially stratified life in these ancient civilizations.

Mining in the Roman World

The depletion of easily accessible deposits of copper and the general scarcity of tin accelerated the thousand-year transition to iron about 3000 BP. One age replaced another in non-linear ways because premodern societies continued to use familiar technologies, despite technological innovations and breakthroughs. For example, the Etruscan civilization of the Italian peninsula reached its pinnacle about 2800 BP and fabricated state-of-the-art iron weapons. Not surprisingly, they also produced beautiful bronze statues and functional copperware. As almost all students of history know, the Roman Republic followed the Etruscans on the peninsula and within a few centuries the Roman Empire dominated the Mediterranean region, referring to the Sea as "mare nostrum," our sea. Its empire spread from the British Isles across Europe and North Africa to the eastern shores of the Mediterranean.

In an effort to support a growing empire, Roman explorers fanned out across the Mediterranean region in search of mineral wealth. In destroying the Phoenician colony of Carthage in North Africa they acquired its Mediterranean possessions, including the ancient mines of Iberia. The best known of these were the Rio Tinto copper mines, near the modern city of Seville. The mines were named "red river" because the residue of copper and iron ore ran into the river from a reddish colored mountain. Excavations in the area recently revealed an amazingly large mining enterprise undertaken by the Romans with the use of slave labor and a highly developed smelting apparatus. Where the mountain once stood, 3,000 years of mining revealed a large crater more than 800 feet deep and over three-quarters of a mile wide.

Intensive mining began at Rio Tinto with the Phoenicians during the Middle Bronze Age and was continued by the Greeks, Carthaginians, and the Iron Age Romans of 2200 BP. Significant mining and smelting took place but the primary focus of Roman mining was the extraction of gold and silver for the manufacture of coins. Beginning in 2200 BP and continuing for four centuries of full-scale production, the Rio Tinto mines delivered about 1,815 tons of silver as well as impressive amounts of copper, iron, and gold. Roman coins became the currency of its growing empire. One might say that the empire was held together and allowed to grow and prosper because of a common currency in silver coins.

The mines contained a near-complete array of "sulfide" minerals, meaning that they were bonded to sulfur chemically. In the copper, iron, lead, and zinc smelting process, the sulfur oxidized and became airborne. The ore that contained very small amounts of silver contained large amounts of lead. In the molten liquid produced by smelting, lead oxide became airborne. By getting rid of the lead, silversmiths molded, hammered, and stamped the metal into silver coins. Given what we know now about the toxicity of lead and the amounts mined during this four-century period, the environmental and health effects must have been significant. By the fall of Rome in 476 CE, the empire had exhausted the more accessible ores.

Estimates of the ancient slag accumulated at Rio Tinto have ranged between 5 and 20 million tons. If one estimated the mid-range of 10 million tons, it would have taken 20 million tons of charcoal to smelt the iron from Rio Tinto's mines. In order to produce that amount of charcoal, 100 million tons of timber would have been consumed. Combined with smelting operations on the island of Cyprus starting around 6600 BP, as well as the area across the Caucasus and Afghanistan, these operations were a major cause of long-term deforestation, the loss of flora and fauna diversity, soil erosion, and microclimate changes in the Mediterranean region.

The Romans engaged in most of their mining and production during the late Republic and early Imperial periods and may have managed the surrounding woodlands by harvesting, pruning, and re-growing. Assuming a 15-year growth period for trees, proper management would allow for the continuous harvesting of woodlands within a 25-mile radius.[7] If this was the case, then we are left trying to explain why production at the mines declined during the later Roman period.

Figure 6.2 Reconstruction of a Roman crane. It could lift up to 7,000 kg and was used in the construction of fortresses and large buildings. Two legionaries or slaves had to walk within its wheel to operate it and it had to be anchored to the ground to prevent movement. *Source*: From Wikimedia Commons, the free media repository.

In an activity that by and large remained small scale and spread out geographically in the ancient world, the large-scale activities at Rio Tinto remain an impressive reminder of the ancient exploitation of nature.

How extensive was the air pollution caused by silver production? Was it restricted to the immediate area, the surrounding region, or the northern hemisphere? Did the Romans know that lead was poisonous? If the air was contaminated, how did that contamination compare to the period from the 1930s through the 1970s in which lead became an additive to gasoline to prevent engine knocking of high-performance automobile engines? The answers to some of these questions were discovered by a group of scientists who began drilling into the two and a half miles of the 400,000-year-old Greenland ice cap to study the Earth's changing climate. Over these many millennia, pollutants mix with water vapor in the atmosphere and fall to earth as snow. With each passing year in the frigid zone of Greenland, snow turns into ice, capturing the pollutants inside the ice cap. The question about atmospheric lead levels in the ancient world and their potential environmental and health effects became a focus of the work of the Australian scientist Kevin Rosman.

During the era of pyramid building in Egypt, little if any lead existed in the atmosphere. As expected, with emerging industrialization in the eighteenth century, lead levels skyrocketed. For the millennia-long manufacturing era, however, what explained the surge of lead levels found in ice cores from 2500 BP until about 900 CE? And how would Rosman determine that the lead came from the Rio Tinto mines? In scientific terms, lead and its isotopes form after billions of years from decaying uranium and thorium. In crustal rocks, the ratio of two isotopes, lead 206 and 207, increases over time. These numbers refer to the number of protons and neutrons in lead nuclei. The lead at Rio Tinto was not trapped in crustal rock but in silver ore and therefore had a lower isotope ratio than lead found elsewhere.[8] So, the lead from Rio Tinto had a unique "fingerprint" and could be located by its location in the ice core. There, lead levels were about four times higher during the period of intense mining and smelting at Rio Tinto between 2150 BP and 50 CE than at any time before this 200-year period.

Not all of the lead evaporated, however, since Roman metal smiths captured much of it for fabrication. Smiths prized the metal for two reasons. They could easily manipulate it and their finished products resisted corrosion more than almost any other known metal. Some identified it as a particularly Roman metal because they used it in plumbing materials, construction, and shipbuilding. In addition, tableware, kitchen utensils, and drinking vessels accounted for many of its uses.

As an ingredient to stop wine from fermenting, lead posed a serious neurological health hazard to unsuspecting Romans. Although we may never know for sure, some historians of Roman life have suggested that lead poisoning contributed to the empire's eventual demise. Atmospheric lead levels remained considerably lower than levels for the period from 1930 to 1970 in which lead levels from ice cores were identified as 25 to 50 times higher than during Roman times. Atmospheric lead levels during Roman times, however, were four to five times higher than at any time during ancient history.[9]

The longer-term environmental effects of mining remain a subject for ongoing research. Ancient mining at Rio Tinto left behind millions of tons of black slag in mounds and hills that transformed the landscape and further polluted the river and groundwater. Silver-lead mining produced large quantities of toxic mining and metallurgical wastes, including zinc, arsenic, and cadmium. They remained in the hills of furnace slag, in ancient waste-rock dumps, in the soil and river sediments. These elements penetrated the entire environment by way of the flowing river, by vegetation growing on the surface of contaminated soil, and by the airborne dispersal of dusts and fumes. Their presence in the environment today, after the Romans terminated mining and smelting operations nearly 2,000 years ago, remains a subject of intense scientific interest.[10]

Rio Tinto fell into disuse after the fall of the empire. As the workforce abandoned the surrounding countryside, Rio Tinto became a ghost town. Once a British mining syndicate purchased the mines from the Spanish government in 1877, it began operations using new smelting technologies for the ore extracted from deeper mines. The mines yielded a brassy yellow copper more difficult to smelt than the green and blue copper closer to the surface. Burning this deeper mined sulfide copper ore released hydrogen dioxide gas. It clogged the air, choked miners, smelters, and smiths, and killed all of the vegetation within a 15-mile radius of the furnaces.[11]

The Age of Iron

After thousands of years of smelting and casting copper/bronze implements and spreading the technologies to cities throughout the Eurasian landmass, iron production emerged among the tribal people of Anatolia in western Turkey who possessed the world's longest tradition of manufacturing metal goods. As metal producers consumed the accessible copper found in shallow mines, experimentation with iron ore became an economic necessity. Ecological considerations also came into play. The smelting of copper and tin placed heavy demands on the environment. As noted earlier, charcoal production led to widespread deforestation throughout the Mediterranean region. Coupled with the expansion of agriculture and the increasing concentration of a growing population in cities, land clearing became a priority. When compared to copper smelting, iron manufacturing using newer technology consumed less charcoal, lowering fuel costs. However, this efficiency got lost in the increased production of iron to meet the material needs of a growing population. As a result, substantial deforestation took place in Greece, Turkey, and Palestine.[12] Soil erosion increased as the forest canopy retreated further and iron smelting contaminated the soil and air in ways similar to the less efficient methods of copper manufacturing.

Although iron did not become an important metal until about 3200 BP, its existence had been known as early as 5000 BP. The circumstances for its discovery in Europe are probably similar to that of China in earlier centuries. As a byproduct from the production of copper, smelters and smiths became intrigued by its metal properties once the easily accessible copper became scarce. They experimented with iron more than a thousand years before it became an economically viable alternative to copper. With its origins in Anatolia and Southwest Asia, iron technology dispersed across some areas of Europe, Asia, and Africa. By 3000 BP, producers of iron used the same basic technology. As we shall see later, metallurgy in Europe, unlike in China, remained small scale, using furnaces that produced wrought iron directly from ore. These small-scale operations were so inefficient that neighboring forests were unable to support many furnaces. European production remained low despite the existence of rich iron ore deposits.

The slow acceptance of manufactured iron was caused partly by the inability of smiths to achieve the high temperatures required during the production process. Iron melts at about 2800 °F, whereas bronze alloy is achieved by melting copper and tin at 1950 °F. Although not considered significant by today's standards, this increase in temperature required energy and technological knowledge unavailable to most early metal smiths. In addition, once iron reached its melting temperature, it became a spongy alloy of iron and carbon that must be separated before the iron can be worked. Learning to remove carbon from this pasty mass required the accumulation of metallurgical and chemical knowledge gained by experimentation and communication with other smiths.

As a soft and malleable metal, smiths made nails, plows, and iron bars by using a small furnace or forge called a *bloomery* to heat the ore to about 1830 °F. Burning charcoal was used as a source of heat to separate iron from its oxygen. Once completed, metal smiths hammered this red-hot spongy mass to remove its impurities, mostly slag. Called the "bloom," it was fashioned then into various shapes. With a carbon content of iron at 5 percent, hard and brittle cast iron was suitable for pots and cauldrons and for tools and utensils. For more than 2,000 years, metal smiths produced small batches of wrought iron in this manner.

The decades and even centuries of trial and error accelerated the learning curve of metal smiths and forecast the arrival of an ever-expanding use of manufactured iron for weapons, tools, utensils, ornaments, building materials, and precious objects. With the invention of more efficient furnaces using less charcoal, smiths were able to achieve the necessary 2800 °F. As a result, iron ore of high quality with low carbon content was forged and hammered into weapons of steel, razors, and axes.[13]

Although the amount of wood consumed by the production of iron was less than that of copper, the increasing demands for iron eclipsed these efficiencies. Charcoal production taxed and depleted local and regional forests. Combating deforestation and soil degradation became severe problems for early manufacturing centers. Fuel shortages limited the production of iron to specific areas and could make iron prices prohibitive for consumers bearing long-distance transportation costs. Eventually, abundant and available coal made it unnecessary to get every ounce of combustible energy from this fuel source.[14] In the process, increasing volumes of smoke, ash, and soot taxed the local environment.

Iron Making in China and India

Iron production in China reflected the growing demand for iron products. By 3300 BP, the Chinese produced their first cast iron. They built larger furnaces to increase temperatures, production, and quality. As a result, they used more wood and depleted more forests. By 2400 BP, however, the use of efficient double-acting bellows with charcoal as the fuel source produced high-quality iron.

Because of their ability to achieve higher furnace temperatures using advanced bellows, the Chinese were casting iron 2,000 years before it was done in Europe. In addition, as a developed pre-industrial economy,

Figure 6.3 Japanese metal workers during its medieval period operating a forge. *Source*: © MEN (Musée d'ethnographie, Neuchâtel, Switzerland). Photo Alain Germond.

China's large furnaces allowed them to burn larger amounts of high-carbon fuel to smelt an equal amount of iron ore. The chemical effect of this larger-scale production process caused carbon monoxide to mingle with the ore and lower the melting point to about 2102 °F, thereby increasing the efficiency of their furnaces.

By 100 CE another technological innovation, the hydraulic bellows, improved production and by 400 CE the Chinese began the transition from charcoal to coal. The iron used in building construction by the Chinese showed the high quality of their iron production, the best example of which was the pagoda at Dong Song in Hebei. It stood 70 feet high and used more than 53 tons of iron.[15] Iron chain suspension bridges increased the flow of commercial traffic and iron plows improved agricultural production. Included in the 125,000 tons of iron produced in China by 1078 CE were 16 million iron arrowheads. Iron weapons modernized China's military apparatus. By the Middle Ages, China's productive capacity exceeded that of Europe.

Only India's technological advances in iron making equaled China's. Evidence of high-quality iron and steel making appears before the common era (CE), by as many as two to three hundred years. Crucible steel appears in the archeological and historical record of India, especially the manufacture of *steel wootz* in 2400 BP, a carbonized wrought iron made in crucibles used to make a sword with sharp edges that were not dulled by hammering. Generically, this process could be considered the origin of manufactured finished steel.

A charcoal crisis and a failure to use manufacturing innovations caused stagnation in India's iron and steel production. The crisis caused its iron manufacturing to atrophy. For centuries this industry's forges had produced arrowheads, daggers, and knives for the armies, and ornaments, utensils, and tools for local populations. About 14.5 tons of charcoal were used in producing one ton of refined wrought iron. To produce the 2,500 tons of this metal would require consuming a surrounding forest area of 437 square miles of woodlands or about 140,000 tons of wood. In India, the persistent use of wood, the absence of higher-temperature furnaces that coal and coke required for iron smelting, and the failure to develop machines to provide powerful blasts of air to replace their low-powered bellows caused stagnation in manufacturing.[16] Despite stagnation, as late as 1750 CE about 10,000 small-scale iron and steel furnaces continued to operate, producing about 20 tons per furnace each year. By then, Britain had begun flooding India's market with less costly iron and textile goods, aiding in the destruction of India's manufacturing capacity.[17]

While iron production in Europe consisted of bloomery smelting in small batches until the medieval period, the Chinese began producing iron in larger quantities by 2300 BP using blast furnace technologies. In Europe, where these technologies had not yet been invented and where iron production was decentralized in small villages near wood supplies, China's Han emperors (2206 BP–120 CE) monopolized control over production and located ironworks near population and administrative centers, not near forests. Chinese technology promoted production on an unprecedented scale and, given its location near administrative/population centers, contaminated urban environments with ash, soot, and smoke. Depleting woodlands led to a fuel crisis everywhere and resulted in the substitution of one fuel source for another and the intervention of new technologies. The revolution in fuel use and new inventions drove the production of iron upward.

The blast furnace is an invention for large-scale, efficient production of at least a thousand pounds of iron each day. This required that the furnace operate for weeks and months without interruption. So, the reliability of fuel and ore, a large labor force, markets for the furnace's output, and transportation became important parts of the production process. The availability of fuel wood posed a constant problem given the insatiable demands of blast furnaces for charcoal. As a result, Chinese iron production in 2300 BP, much like the ironworks in colonial America in 1700 CE, required the organization of production in *iron plantations* either near large tracts of forest for charcoal production or along transportation networks that made the woodlands accessible. With the depletion of these woodlands, iron production facilities were moved to locations closer to fuel sources.[18]

The production process consisted of charging the blast with iron ore, charcoal (fuel), and flux (normally limestone) that entered from a shaft in the top of the furnace. A blast of air was blown continuously into openings near the bottom of the furnace. Iron and slag were tapped repeatedly from another opening at the base. Given the size and construction of these furnaces, they remained productive for years and produced cast iron fashioned into useful products or converted to wrought iron. They also generated intense heat and released hot plumes of air into the atmosphere.

The furnace at Guangzhen in Long County, Henan during the Song dynasty (1040 CE–721 CE) was 6½ feet thick. Its 19½-foot base of compressed earth contained high concentrations of fine sand, powdered charcoal, and powdered iron ore reinforced with a retaining wall of tamped red clay as much as 30 feet thick. Although there are no reliable ways to calculate the large-scale environmental effects of these furnaces, air quality deteriorated, hills of slag contaminated the soil and groundwater, and the dangers to human health from heat and fire as well as from breathing the fine red metallic dust must have been life threatening. Since the bellows that provided the blast of air was generated by human labor rather than water or steam power, the demands on the workers remained constant. Air pollution of this sort was a harbinger of the air pollution of iron- and steel-making industrial plants in the modern era. Workers labored under conditions of heat and trauma unknown by previous generations.

The Chinese began using coal about 1100 CE while Europeans, particularly the British, began using it about 1700 CE. Although the transition from charcoal to coal was separated by many centuries, the depletion of woodlands and the resulting fuel shortage necessitated change in both places. To begin the process, Chinese workers broke up iron ore into two-inch chunks and used sieves to eliminate smaller pieces. As a result, ironwork sites contained hills of mined but unused ore. Although these sites must have been an eyesore, iron masters knew through experimentation that using two-inch fragments of iron required a greater blast pressure. Since a greater blast necessitated more fuel and more human labor, the decision was a tradeoff between economies of scale and fuel consumption on the one hand and fouling of the landscape surrounding the furnace sites within the Chinese administrative/urban centers on the other.

In addition to hills of unused iron ore, the iron production process revealed some additional environmental costs. The ore at these large-scale Chinese ironworks contained about 50 percent iron. The remains consisted of slag, silica (SiO_2) in the slag, and lime (CaO), largely from the ore and the limestone in the charge. About two tons of ore were needed to produce one ton of iron. There is no way to estimate the amount of iron produced daily or annually, but using Chinese blast furnaces in operation during the past two centuries, a figure of 100 tons per year for each of these large-scale ironworks seems plausible. This estimate would suggest national production during the Han dynasty at 5,000 tons annually or about 0.22 lbs per person.[19] The famous poem by Su Shi in 1079 CE titled "Stone Coal" aptly captured the impact of the blast furnace on the air quality of the local environment and the concentration of the labor force at these large scale furnaces:

> The stinking blast – *zhenzhen* – disperses
> Once a beginning is made, [production] is vast without limit.
> Ten thousand men exert themselves, a thousand supervise.[20]

By 1000 CE, water power had begun to replace manpower in operating the bellows that created the blast and in an effort to conserve the country's woodlands, coal and coke began to replace charcoal as a fuel source in China's well-developed manufacturing sector. Wrought iron and cast iron were produced in large enough quantities to satisfy the growing commercial and consumer sectors of the economy.

Iron Making in pre-Modern Europe

The assessment that metallurgy and manufacturing in Europe remained relatively unchanged for centuries after the fall of Rome is an error perpetuated by the use of terminology such as the "Dark Ages" or "Middle Ages." These "ages" book-ended the period in European history separated by the fall of the Western Roman Empire in 476 CE and the sacking of Constantinople by the Ottoman Turks in 1453 CE, bringing an end to the Eastern Roman Empire. The centuries were marked by momentous technological changes in agriculture, in waterpower, in building construction, in the manufacture of textiles, in metallurgy, and in weaponry. The changes were gradual and in most instances incremental improvements on methods of production and on the organization of work.

In the immediate aftermath of the events in 476 CE, life along the broad Mediterranean seaboard remained vibrant despite the chaos created by invasions from Germanic tribes from the north. In the north, however, population densities remained very low. The population of Gaul (modern France) in 1500 CE has been estimated at fewer than three persons per mile and for Germany and Britain less than one person per mile. Despite low density and shorter life spans caused by malnutrition and disease, the north possessed ample forests, accessible metal ores, and numerous rivers and streams. Coincidentally, climate events probably accounted for most of the backwardness of the north. Glacial ice penetrated northern Europe in the fifth century and did not retreat until the middle of the eighth. Only with the retreating ice did population growth and economic development come to the northern regions of Europe.

By 850 CE, the towns of Southampton and Ipswich in England had built settlements designed in a grid pattern of workshops, stalls, and storehouses. Wines, glassware, and wheel-thrown pottery were traded for hides and wool. The workshops forged the innovative heavy plow mounted on wheels connected by a fabricated iron axle and pulled by oxen. The wheels allowed farmers to adjust the plowshare to the depth of the furrow. Multiple harnesses (made from hides) allowed one ox or horse to be connected to the next or behind another for the labor-intensive burden of removing trees and plowing fertile lands. Along with the invention of the harrow, the scythe, and the pitchfork, improved axes and spades revolutionized agriculture and increased food production. Iron tools made by metal smiths, as well as harnesses, padded horse collars, and breast straps made in the workshops of tanners, released humans from the challenges of manual labor.[21]

A second and equally significant innovation occurred in weaponry and improved the fortunes of armored horsemen who ruled Europe throughout most of the Middle Ages (500 to 1400 CE). Existing battle-axes, the new long sword and heavy mace, iron-tipped spears, lances, helmets, and chain metal in the hands of experienced warriors and medieval knights transformed conditions on the battlefield. After the fall of Rome and the disappearance of state sponsorship, the production of iron required a diffusion of skills employing skilled metal smiths, apprentices, haulers of fuel, and laborers working in local forges. Their presence was felt everywhere, and the elevated status of some craftsmen placed them next to rulers and royalty.

Despite the wave of decentralization, however, manufacturing also exhibited some impressive examples of concentrated iron production. For example, a 24-furnace complex located at Zelechovice (now in the Czech Republic) employed thousands of workers in logging, mining, production, and fabrication during the Middle Ages. Although the exact dates of production remain in doubt, the archeological remains suggest a facility of impressive proportions.[22]

Productivity during the Middle Ages witnessed the growth of agriculture, cities, and manufacturing. Between the Baltic and the Mediterranean Seas, European populations grew from about 27 million in 700 CE to over 70 million in 1300 CE.[23] The majority of the population continued to toil on the land as tenants but many were freed from the shackles of slavery and serfdom that dominated life in antiquity, as rural towns and growing cities became centers of activity and production. In the towns and cities, manufacturing flourished. Small, decentralized, family-oriented, cottage manufacturing provided employment and emancipation for the rural masses. In addition to the iron goods identified earlier, the manufacture of hoes, sickles, hooks, saws, hammers, sledgehammers, drills, knives, tongs, files, nails, picks, chisels, tie-rods, and much more became commonplace in an environment dedicated to material progress.

Figure 6.4 Steel production in the Middle Ages. *Source*: From Wikimedia Commons, the free media repository.

To accelerate the rate of material progress, a technological breakthrough was achieved during the Middle Ages with the use of the blast furnace combined with the power output of the mechanical waterwheel. As noted earlier, blast furnaces operated in China and Southwest Asia long before they either arrived in Europe or were invented independently by Europeans. The first one excavated in Europe was located in Lapphytten, Sweden, possibly functioning before 1350 CE. Waterwheels mechanically pumped a concentrated blast of air through bellows to a common opening in the furnace. The stronger blast quickly raised the ore's temperature to the point where carbon combined with iron rapidly. This innovation produced an alloy of 4 percent carbon and 96 percent iron. It had a lower melting point of 2012 °F rather than the 2780 °F required to melt iron ore in a kiln.

More efficient production followed technical improvements in waterwheel technology.

The replacement of human labor with mechanical power transformed the productive capacity of medieval towns and cities located along waterways. By 1200 CE, watermills had wheels three to nine feet in diameter improving power outputs from one to three horsepower. In 1600 CE, builders fabricated 32-feet diameter wheels that resulted in a corresponding improvement in horsepower. As is true with many technological innovations, implementation did not follow rapidly on the heels of invention. Many mill owners preferred the small-scale wheels to avoid the complications involved in concentrating energy on a single wheel. Small waterwheels performed a number of functions: grinding grain, making copper pots and weapons, sharpening instruments, polishing armaments, sawing wooden planks, and spinning wool, cotton, and silk.[24] The processing of wool, cotton, and silk dominated expanding textile manufacturing, as waterwheels became the source of power for looms.

At the beginning of this period, Europe's population growth exceeded its agricultural capacity. Rents increased while real wages declined. In succeeding decades, however, the continent's population crashed. A devastating famine gripped northern Europe in 1317 CE, followed by an equally disastrous one in the south from 1346 to 1347 CE. The Bubonic Plague decimated Europe's population from 1348 to 1351 CE. From an estimated continental population of about 80 million, close to 20 million died in a little more than two years. Wars overlapped these natural events. The Hundred Years' War (1337–1453 CE) destroyed vast agricultural lands, villages, and towns in France. In England, the War of the Roses (1455–1485 CE) pitted countrymen and women against each other and brought misery to both sides. England's population dropped by one-third as a result of wars, famine, and disease but, similar to life on the continent, the country's manufacturing per capita grew by 30 percent. In 1400 CE, England's iron production was about 30,000 tons. A century later it exceeded 40,000 tons with a significantly smaller population.[25]

Cast iron replaced wrought iron, revolutionizing manufacturing during the Middle Ages, and foreshadowed the industrial age. Water-powered blast furnaces required less human labor and ran continuously as long as the water in the rivers and streams continued to flow. Damming these watercourses and holding water upstream eliminated seasonal variations and allowed manufacturing to continue regardless of season, climate, or weather patterns.

In this newly energized world of metallurgy, smiths became even more visible members of medieval towns and cities. Their importance probably increased still further with the impact of such economic factors as improved transportation, increased trade, and more reliable markets. By 1400 CE, blast furnaces were operating throughout Europe, from the Rhine Valley to Sweden, Austria, and Belgium. The continuous output of these furnaces provided metal for a smith's forge and, with the expansion of the craft, neighbors began to complain of the nuisance in noise, smoke, and odor. In 1377 CE neighbors of a London armor maker sued him, complaining that:

> the blows of the sledge-hammer when the great pieces of iron … are being wrought into … armor, shake the stone and earthen party walls of the plaintiffs' house so that they are in danger of collapsing, and disturb the rest of the plaintiffs … and spoil the wine and ale in their cellar, and the stench of the smoke from the sea-coal used in the forge penetrates their hall and chambers.[26]

This suit provides a fourteenth-century example of those that would be lodged by city inhabitants with increasing frequency in the centuries to follow. They viewed these intrusions into their daily lives as nuisances that offended their senses. Not until much later in modern history would these nuisances become viewed as contaminants, pollutants, and despoilers that damaged the natural environment and compromised human health.

To accelerate the production of iron goods and to provide metal smiths with a steady and increasing supply of iron for their forges, innovations were required to improve production processes and to reduce the amount of human labor required in the fabrication of ironware. In the 1450s, a trip-hammer was applied to pound wrought iron into suitable shapes. Having been used in China for centuries, it consisted of a heavy iron head attached to a long wooden shaft "lifted and released by a drum armed with cams. Rising, it struck a wooden spring beam; the spring's recoil added force to the down stroke. Alternately an iron block in the floor under the hammer tail achieved the same result."[27] By 1500 CE, the demand for nails for construction exceeded the supply and the invention of a slitting mill provided the smith with long slender iron rods that could be easily cut into nails. This ingenious invention consisted of a pair of rotary disk cutters turning in opposite directions. As one writer described it: "The first piece of true machinery after the power hammer to be introduced ... of even greater importance, it contained the elements of the rolling mill."[28] In the age of industrialization to follow, the refinement of the technology of the rolling mill would be essential for the mature iron industry and a fledgling steel industry.

By 1500 CE, European production of iron reached 60,000 tons as machines came to play an increasingly important role in the making of goods for the home market and for export to other countries. Water power, as we have seen, continued to penetrate all sectors of the economy, from the hydro-powered sawmill, to the grist mill for making flour, to the textile mills for making clothing and, of course, to the iron foundry for making machine and craft tools, household utensils, and weaponry. These technological advances came at a price, however. To provide for a growing population and the materials to build towns and cities, forests were cut to provide timber for a seemingly unending array of new and improved goods. These included construction materials, furniture, cooking in an oven, heating the forge and blast furnace, tile and brick making, and glassmaking. The ecological footprint of these human activities placed added stress on the quality of the land, water, and air. Land was cleared for the production of food and consumer goods, while water-powered mills discharged their wastes into the streams and rivers. New manufacturing facilities continued their assault on air quality.

If the period from 1500 to 1600 CE established the conditions for industrialization, it also created the conditions for an energy crisis. The Age of Transatlantic Exploration from 1500 CE onward required a boom in shipbuilding, in weaponry, and in metallurgy. The consumption of central Europe's remaining forests increased considerably. By 1550 CE, the smelting of silver from the German Fribourg mines consumed 2.1 million cubic feet of timber for fuel each year. Two other mines in north central and northwestern Europe, at Huttenberg and Joachimstal, consumed an equivalent amount. In the same region, the districts of Schlaggenwald and Schonfeld used up 2.6 million cubic feet each year.

The forests of the region literally disappeared within a generation. With such a dramatic depletion of woodlands, the costs of fuel skyrocketed, representing as much as 70 percent of the production costs of iron in some districts. Foundries closed because of the exploding fuel costs; others curtailed production, while yet others witnessed a general economic decline caused by rising fuel costs. In Genoa, oak used in ship construction grew from a base price of 100 to 1,200 lire in 120 years. By 1600 CE, the Italian economy had begun a period of long-term decline.[29]

Manufacturing in Colonial America

Despite extensive deforestation after millennia of exploitation in Europe and Southwest Asia, in much of the rest of the world, including North America, wood remained the most important renewable material. Compared to mining, it was cheap to harvest and remained easier to shape and fasten into objects and equipment used in the home, the farm, and the artisan's workshop. In the United States until 1850 CE, wood was the primary household and manufacturing fuel source.

In transportation, burning wood provided steam for the boilers of steamboats and locomotives. It was used more than coal for smelting, refining, and forging metals. Wood was the ubiquitous building material. The construction trades built timber-framed houses, waterwheels, watermills, windmills, bridges, and military installations. Ships were built with wood. The wine industry required wood for its casts and vats. In the tanning of leather goods, the bark from trees was used as a dye. Rope makers used wood fibers. Glassblowers fueled their furnaces with timber. Almost all machinery was made of wood.

Deforestation in colonial America was accompanied by all of the attendant environmental costs, changes in microclimates with extremes in temperatures becoming commonplace. Soil erosion, runoff and flooding, the silting of streams and estuaries, and the clogging of main waterways became breeding areas of malaria-carrying mosquitoes. Next to the forest, the farm became the source of raw materials for manufacturing. Fibers from hemp became textiles, while tanned hides from animals became leather goods and belts for the wheels of small manufactories. Slaughtered livestock provided lubricants of lard oil and tallow for mechanics and millwrights.

Many of the manufacturing activities described here damaged the natural environment, but the impact on the health of men and women must have been equally severe. For miners, lung disease and its related maladies were serious illnesses. For gilders of gold, the process of amalgamating gold and mercury and then evaporating the mercury with fire poisoned their lungs and led to an early death. Potters used pulverized lead to varnish their earthenware. They suffered from a host of illnesses, from the loss of teeth to paralysis, lung disease, and eventual death.[30]

CHAPTER SEVEN

INDUSTRIAL WORK

Introduction

More and more rural people combined agriculture and manufacturing to produce textiles, tin, bronze and ironware, and a range of other products for household use. These rural handicrafts have been described as "proto-industrial" and as the beginning stage of modern industrialization.[1] For many generations, the growing importance of manufacturing in world history overlapped the rise of industry.

Throughout the "proto-industrial world," homes produced manufactured goods for local consumption. Craft shops owned and operated by artisans grew to satisfy the local demand for dry goods, leather, furniture, farm and household tools, and utensils as well as precision instruments. This long-term historical development identified as manufacturing and the transitional "proto-industrial" phase ultimately became a commercialized society in which consumer demands shaped a newly emerging market economy – an economy in which "all production and consumption became totally oriented toward selling and buying in the marketplace and that everything – goods, land, labor, even time – became valued accordingly by the calculus of supply and demand and the cash nexus."[2]

China and India's Economy

The genesis of this new calculus emerged in China and India centuries before it blossomed in Europe, for as recently as 200 to 250 years ago, these two Asian countries accounted for much of the world's economic productivity. From 1500 to at least 1700 CE, Europeans purchased increasing quantities of Asian silks, colored cotton textiles, spices, and porcelain from Indian and Chinese producers using the large supplies of silver plundered and extracted from Spain's newly conquered colonies in the Americas. Used by the Spanish kings, Charles V (son of King Ferdinand and Queen Isabella) and his successor, Philip II, to finance their many wars to unite Europe under Spanish hegemony, they squandered this newfound wealth and became indebted to Dutch, English, and Italian bankers and arms suppliers. When China replaced paper currency with silver coins to stabilize its monetary system, the world's most productive economy now had access to the monetary engine that provided for its continuing growth. About 75 percent of the world's growing supply of silver enhanced the purchasing power of European traders in China from about 1500 to 1800 CE.

By 1775 CE, Asia, but primarily China and India, produced about 80 percent of the world's goods and it seemed unlikely at the time that either Europe or their colonies in the New World would be able to compete in either quality or price with these Asian behemoths. India produced about 25 percent of the total, mostly in lightweight, colored, finely woven cotton cloth. These "calicos" were delicate to touch, comfortable against the skin, and cheap by European standards. The prospect of competition from either Europe or elsewhere seemed fanciful.

A number of historical events turned the tide in favor of European (mostly British) fledgling textile industries that were primarily supplying local markets. New World supplies of cotton, grown on southern plantations using ever-increasing numbers of African slaves, fed incipient British textile mills. In addition to supplying local British markets, the output of their mills in Manchester and elsewhere supplied cheap cotton goods to overseas markets that included clothing for plantation slaves. The invention of a mechanical cotton gin by Eli Whitney removed seeds from short-staple cotton and gave a major boost to cotton mills in Britain and the United States. Tariffs imposed on foreign textile goods made them more expensive than local goods. The invention of steam-powered machines in Britain used the stored energy of fossil coal and water-powered mills in the United States further tipped the balance of trade toward Europe and the United States. Europe's aggressive military interventions caused India's declining Mughal Empire to disintegrate further, leading to its eventual conquest and incorporation into the British imperial system by 1850 CE.[3]

China's position as the world's economic powerhouse faced many pressures from Europe's incipient industries. The transition to a fossil fuel economy replaced human labor with machines, while a European tariff system undercut Chinese high-quality, low-cost consumer goods. A global trading network fed by China's demand for silver and Europe's ability to feed that demand opened Asia to an enhanced European presence, thereby gaining access to an Asian world of high-quality commodities. In addition to the demand for silver, Britain's intervention in China's internal affairs fostered the spread of an opium addiction among China's population. China's 40 million opium addicts in the 1800s needed a ready supply of this addictive substance and with Britain's control over the coastal city of Hong Kong they found a willing supplier. Britain's defeat of China in the first Opium War (1839–1842 CE) opened the way for a systematic trade in drugs for the next 20 years. About 50,000 chests of opium, amounting to 6.5 million pounds, entered China each year for sale.[4] As silver flowed out of China to pay for opium grown in India and Turkey, addiction spread. With the British imposing their will on the Chinese government, it forced the legalization of opium and added a number of port cities to the list of suppliers. To benefit from the traffic in opium, Chinese farmers began to replace cotton with poppy fields, increasing their cash flow. This decision made them dependent on others for their food and as a result made them vulnerable to crop failures and famine. By the end of the nineteenth century, China's population was consuming 95 percent of the world's opium supply, with India becoming a major supplier. The impact of the drug trade on both countries contributed to their demise as the world's greatest manufacturing centers in the centuries before industrialization in the West.[5]

European Hegemony and British Industrialization

Britain's emergence as the world's largest export economy in the nineteenth century catapulted it into a position of industrial pre-eminence. As the "workshop of the world" its economic transformation represented an important breakthrough in the relationship between humans and the natural world. That breakthrough was fostered by the human conquest of the material world, probably the most important development in the past 400 years.

As noted earlier, the medieval period in European history introduced machinery on a scale broader than anywhere else in the world using waterpower. It was used first in rural economies where fulling, a process of cleaning and thickening of cotton cloth, enlisted water from the many rivers and streams in the countryside to complete the process. The expansion of this cotton industry into wool, linen, and silk set the early pace for British industrialization. As late as 1770 CE, the British produced little cotton and imported most of their cloth from India. Yet, 70 years later, Britain's productive capacity had grown by over 1,000 percent and the British had become global exporters of cotton. For this to happen, a few very important changes occurred in rapid succession.

A factory-based economy to produce cotton thread needed to replace the rural household. Britain needed large supplies of raw cotton from its slave-based colonies and the technologies for ginning and spinning such large quantities. Owners required capital to construct factories, a large labor pool to work the new machines, and a global market for their cloth. As economic historian Joel Mokyr has pointed out, "In the early stages of the Industrial Revolution, the fixed cost requirements to set up a minimum-sized firm were modest and could be financed from the profits accumulated at the artisan level."[6] Few countries in either Europe or the rest of the world were able to fulfill all of these conditions with the speed of British operators. So, some of the focus of this chapter will center on industrial developments in Britain with a recognition that industrialization spread across Europe and the United States without regard to national boundaries in the nineteenth century but stagnated in both India and China, the world's greatest export economies in previous centuries.

Economic Developments in China, Japan, and India

China, where cotton cultivation and spinning originated, met all of the criteria for entering the industrial age, with one exception. They made the technological transition from muscle-powered looms in the 1300s to water-driven textile machines. Before this breakthrough, spinning cotton and reeling silk were performed by multi-spindle machinery powered by foot pedals. China's extensive system of waterways connected cotton growers with internal markets composed of tens of millions of consumers by the early 1500s. The country's global leadership in producing luxury ceramics, silks, and porcelains was buttressed by private capital investments in a labor-intensive complex of factories. In the process, it created a wealthy merchant class whose accumulated assets exceeded that of many wealthy merchants in Europe and the Americas.

Why, then, was there no rapid industrialization in China? Historians have answered this question in a number of ways. World historians challenge explanations that focus on western exceptionalism. Others claim that China's surplus population during industrialization served as a barrier to gaining production efficiencies through a factory system. Low-cost household labor would continue to produce cotton more cheaply than any factory could and as a result prevented capital investment in cotton thread and cloth. Comparing China's population growth with that of Europe during similar periods, however, shows no population surpluses. From 1500 to 1700 CE China's population grew by 45 percent and Europe's by 48 percent. For a longer period, 1500 to 1900 CE, China's population grew by 330 percent and Europe's by 380 percent.[7] As historian Jack Goldstone has noted about these data and the excess population hypothesis, "Our vision of China as massively overcrowded dates more from the twentieth century, when Asian population growth accelerated while European population growth dramatically slowed, than from any consistent historical differences in growth rates, and our projection of such views onto early modern China is an anachronism."[8] Rather than becoming a barrier, population doubled from 6.4 to 13.1 million in England over the course of two generations from 1770 to 1830 CE, with industrialization accelerating at the same time.[9]

Goldstone has offered a thought-provoking explanation for China's failure to adopt the factory model for production. From the beginning of the Ming dynasty (1368–1644 CE), men tended the fields, cultivating crops and growing, cleaning, and drying cotton. Women were restricted to the household, spinning and weaving cotton cloth and preparing food. A spatial and functional segregation existed between men and women, establishing hierarchical social norms and a rigid moral code. Even as late as the twentieth century, recruiting women to work in cotton mills met with resistance and failure. There was one exception: 72.9 percent of the factory workers in the Japanese-owned mills in Shanghai were women; 26 of China's 34 cotton mills were located there in 1897.

As late as the 1930s, fewer than 10 percent of Chinese women worked in the other mills. A long history of restricted mobility for women and a traditional household economy in spinning and weaving cotton served as an insurmountable barrier to technological innovation and a centralized factory system.

Although women faced an unequal legal and domestic status in Europe, they experienced life differently than women in China. In the proto-industrial European household, textile manufacturing was divided by gender. Mothers and daughters spun, fathers and sons wove, and all farmed as a single unit household. As demand for textiles increased, many families committed more resources to spinning, weaving, and leaving the household to engage in paid wage labor. Men and women alike migrated to towns and cities to work. Married men brought their wives and children to work with them in these non-household forms of employment. Factory owners recruited women and youth (13–18 years old) because of the lower wage scale for them, when compared to males.

These low costs put them at a competitive advantage with Europe's household economies in textiles. But because of restrictive codes governing the role and mobility of women in society, China refused to make the leap of replacing women's household labor with factories and the Manchu Qing dynasty (1644–1911 CE) reinforced this tradition despite the country's history of inventing a multi-spindle spinning wheel and the drawloom for cloth and silk production as early as 1313 CE (454 years before the invention of the spinning jenny in Europe).[10] In addition, China failed to transfer its capital investment strategy and factory system to the production of water-powered cotton spinning mills. And it resisted technological improvements in textile production that became known throughout the industrializing world, including innovations adopted in the production process by Japan.

Japan embraced advances in textile technology and the employment of women in factories compared favorably with Europe. Ruled by the Tokugawa shoguns (1603–1867 CE) that isolated Japan from the West but not from China or Korea, women were allowed a degree of freedom uncommon in China. They moved freely, left their families to work for wages, selected their own marriage partners, and exercised life choices without severe restrictions. In the decade before the end of the Tokugawa rule, typical Japanese villages experienced the mobility of three-fourths of its female population working elsewhere. With the collapse of shogun power after 1867 CE, Japan moved quickly to industrialize on a European model. "Young unmarried women predominated among cotton textile workers" and "migration of farm women seeking employment ... was not a new phenomenon [but built] on a long tradition of female migration"[11]

The path of manufacturing and industrial development in India differed markedly from that of its Asian neighbors. Before machine spinning and weaving in Britain in the second half of the eighteenth century, India's textile production dominated all global markets in Asia, Africa, and Europe. Many countries using local supplies of raw materials possessed some capacity to clothe their indigenous populations. Finer cloth and luxury textiles came from India, however. Foreign observers noted this advantage and tried unsuccessfully, until the invention of industrial machines, to compete with Indian products. Before the adoption of these machines, specialized technical skills in the preparation and treatment of natural fibers and India's comparative advantage in labor costs deterred entry into textile manufacturing.

British and French weavers often imported Indian thread for their domestic cloth production. Over the centuries, Indian spinners and weavers had achieved an empirical and hereditary knowledge of dye-fixing techniques that allowed them to figuratively paint cloth with multi-colored patterns. They discovered that certain kinds of cloth such as muslin could be spun only on a spindle and not on a spinning wheel. And finally, much Indian cloth proved to be more durable than its machine-made competition.

During the first half of the eighteenth century, India reached a peak in textile exports but by the end of the century its overseas markets had been disrupted and its manufacturing capacity for fine cotton thread and cloth was in decline. In eighteenth-century manufacturing, India, the world's largest producer of pure cotton calico textiles, satisfied the world's demand with tens of thousands of rural weavers using hand-looms to supply the growing global market. This highly diffuse manufacturing capacity left a light footprint on the land. Neither the dyeing, bleaching, nor cleaning processes that occurred in thousands of households, and therefore were not concentrated in any one place, contaminated the land or streams.

Industrialization in Britain and the production of cheap machine-made textiles in the nineteenth century, however, severely compromised India's global competitive textile advantage. Trade barriers that put

Indian textiles at a price disadvantage in the overseas trade weakened Indian household and urban weavers to the point that India for the first time became a net importer of foreign cotton cloth. Although Indian producers failed to industrialize their manufacturing in light of the increased demand for cloth from the West, increases in output were achieved by expanding the labor pool while the technology of textile production remained unchanged. To survive, Indian weavers concentrated on the production of coarse cloth for local and domestic markets.[12]

British manufacturers who felt threatened by this competition lobbied the government for protection. Succumbing to the demands of the woolen industry, the British government banned the importation of Indian textiles in 1700 CE, thereby eliminating a global competitor for its domestic market.

Britain's defeat of France in the Seven Years War (1756–1763 CE), a truly global war fought in Europe, the North American colonies (named the French and Indian War), the Philippines, and in the Indian subcontinent eliminated the French who had maintained factories in a number of India's major towns and villages.

Harnessing the Power of Water

In Europe, the harnessing of steam power revolutionized industrial production. Steam engines represented the most advanced industrial technology of its time. In Belgium in 1830 CE, 21 percent of the 428 steam engines were foreign built but 14 years later only 7 percent of the 1,606 engines were manufactured elsewhere. The country also exported steam engines seven times as much in value as she imported.[13] Likewise French textiles, metals, and engineering developed niches in a Europe dominated by Britain. The British produced more iron but the French exported large amounts of finished metal goods to other European countries. The French manufactured higher-quality cotton fabrics competing with cotton produced by dominant British mills. A dynamic equilibrium existed in a Europe dominated by the productive capacity of Britain's superior machinery. When German labor costs were lower and machine technology was equal, British yarn was manufactured on German looms and British iron produced German metal consumer goods.[14]

With New World colonies providing raw material and a market for finished textile products, early cotton manufacturers combined linen and cotton to make "fustian," a cheaper cotton for domestic and export markets. Initially, demand at home exceeded supply but the proliferation of many rural textile manufacturers and the expansion of the Manchester mills created a more balanced exchange. The link between supply and demand, raw cotton and finished textiles, colonial markets and modern producers led to a vast expansion of the industrial sector in Britain by 1750 CE. By then, cotton exports to Africa and to the North American colonies expanded greatly. The cotton crops of slave plantations in the southern colonies became closely linked economically to the cotton textile mills of Manchester, England. "The most modern centre of production thus preserved and extended the most primitive form of exploitation."[15]

To keep pace with demand, inventions including the "water frame," using spindles and rollers, and the addition of handlooms and manual weavers kept pace with the demand for finished goods. As a result, industrialization spread out to country places with great waterfalls and its footprint, although growing, remained in harmony with the ecology of the countryside and the emerging cities. By the end of the eighteenth century, Europeans operated more than 500,000 watermills, with many of them using more than one wheel.[16] The hydraulic energy of moving water remained the main source of power for the mills throughout the eighteenth century and as late as 1838 CE it still supplied one-quarter of the nation's power.[17]

The impression created by novelists and social critics of industrialization in Britain has it that massive factories oppressed hundreds and even thousands of women and child workers. In cotton textiles, where the scale of the system was larger than in wool, linen, and silk, many small enterprises dominated the industry past mid-century. It was the establishment of a capitalist system of middlemen who controlled monetary credit, the supply of raw materials, and the distribution of finished goods that facilitated a revolution in the production of textiles. This new class of capitalists demanded a quickened pace of production and improved quality of goods. In the absence of the factory, significant changes in work were under way. Competition, pressure from larger workshops on smaller enterprises, challenged irregular patterns of work. These conditions, however, transitional to a fully developed factory system, established the pathway to industrialization.[18] Once again, the ecological footprint of early industry remained compatible with the surrounding environment, including its waters.

Decentralized manufacturing and mining employed from a quarter to one-third of the workers in Britain in 1700 CE. With the addition of textiles, and for the rest of the century, a shift in the relationship between agriculture and industry occurred, with industrial work and mining increasing their share of the economy. By 1800 CE, agriculture's employment share declined to 36 percent and a century later to 8.7 percent of the labor force.[19] With manufacturing, mining, and a growing mechanized factory form of industry dominating Britain's economy, the emergence of factory towns with their creaky urban infrastructure of roads, buildings, water supply and disposal systems became more apparent. Road construction and railroad expansion signaled the beginning of industrial capitalism. Canals, railroads, factories, and steam engines became symbolic of industrializing Britain.

Some manufacturing cities in England grew by 40 percent, led by Leeds, Birmingham, and Sheffield in the north from 1821 to 1831 CE. From 1801 to 1851 CE, Liverpool grew from 77,000 to 400,000 and Birmingham grew from 73,000 to 250,000. Manchester, the pre-eminent industrial textile mill town in the nineteenth century, expanded greatly from 1772 CE with a population of 25,000 to 1850 CE with 367,232.[20] Europe's largest city, London, also continued to grow but at a much slower rate. In 1801 CE, 40 percent of those living in cities with 5,000 residents or more were living in London. Fewer than one-third lived there in 1851 CE. With a population of two and a quarter million people and the country's largest port, it was six times larger than its closest rival, Liverpool.[21]

The strain on the built environment became visible with the influx of dispossessed agricultural workers, attracting mostly young men, women, and children. Building owners divided and subdivided again existing living quarters. Cellars and garrets became crammed with occupants, with many families sharing the same rooms. Wages were low but rents soared as price gauging by building owners became commonplace. Poor housing, faulty sanitation, inadequate nutrition, crime, disease, and poverty took their toll on a population where work, any kind of work, even dangerous work, served as a magnet to these newcomers.

Disease, Death and a Public Health Response

Diseases, primarily tuberculosis but also typhoid and cholera, spread rapidly, causing periodic epidemics. These slowed economic growth and energized industrial groups to become active in promoting urban hygiene. The nightly washing of streets of manure and urine, the covering of open sewers with culverts, and establishing local boards of health to impose housing standards on owners became municipal initiatives. Vaccinations for all residents and the expansion of public hospitals were all intended to reduce mortality rates and reduce the spread of airborne and waterborne diseases. By the 1830s, considerable progress in public hygiene had been made, as births began to exceed deaths in many of these cities. Recognition did not necessarily result in the enforcement of changed public health policies but some changes did take place.

Late eighteenth-century plans to cover open sewers in these cities stalled with the expansive growth in population. The proximity of drinking water wells to privies and sewers posed a constant health hazard. Sewage disposal remained a primary municipal dilemma. Labor shortages, more than the loss of credit and capital, stymied economic growth, so a regular flow of workers into the cities to fuel growth in production became a necessity.

Given the metabolism of growing cities, energy produced by the consumption of food produced human and animal waste. This natural and organic phenomenon would plague industrial cities into the modern era as they struggled in search of the ultimate sink.[22] In the interim, rivers, streams, ponds, and lakes became sinks polluted with human waste, animal carcasses, dyes from mills and tanneries, pulp from paper mills, and sawdust from lumber mills. Filth and smells unknown in sanitized modern cities were ever-present.

Numerous cellar flats, housing 12 percent of Manchester's population, made it the most visible example of decrepit living standards. Tuberculosis spread with sub-standard and crowded conditions and a diet deficient in calories and protein further weakened cellar dwellers, making them vulnerable to infection. To lessen the threats, Manchester attempted to drain its privies and cellars in 1845 CE by removing 70,000 tons of excrement from the city. The challenge overwhelmed the municipality, however, because the 38,000 privies outnumbered toilets connected to a primitive sewer system by four to one in 1869 CE.[23] Many of Manchester's residents lived in a "germ factory," where fecal contamination infiltrated the food chain and the water supply. Diarrhea, dysentery, cholera, and typhoid attacked the young, the infirm, and the elderly. The natural world's foundation resources, namely the land, water, and air, carried microbes lethal to human health and survival.

In 1800 CE, London was the only city in England with a population of 100,000 but 50 years later nine cities achieved that status. All were centers of industry. Manchester, Liverpool, Birmingham, and Leeds were remarkable for their explosive growth, industrial activity, and legendary urban fifth and squalor. Charles Dickens, Britain's celebrated novelist and critic of industrialization, wrote about his fictional Coketown as an urban place "where Nature was as strongly bricked out as killing airs and gasses were bricked in."[24]

Britain's industrial prowess was further stimulated by the transition to steam from waterpower and the use of fossil coal to convert water into steam. One of the more visible manifestations of this technological transfer was the appearance of as many as 500 factory chimneys in Manchester belching dense, black smoke composed of coal tar, soot, and ash. As air quality plummeted, respiratory diseases rocketed upward.[25]

Sustained efforts to clean up the physical environment and promote human health remained uneven throughout the nineteenth century. Although other industrial mill towns such as Birmingham and Liverpool provided public baths and washhouses in order to promote hygiene for the poor, Manchester had none in 1865 CE. Slum clearance, a priority in other cities, was not addressed in Manchester before 1891 CE. To its credit, however, the city banned the construction of back-to-back houses in 1845 CE to relieve overcrowding. It appointed a medical health officer in 1868 CE, attempting to slow the spread of diseases which, in addition to those noted earlier, included chicken pox, whooping cough, and scarlet fever.[26] In 1826 CE, the average English life expectancy at birth reached 41.3 years, one year and four months shorter than the 42.7 years achieved during Shakespeare's lifetime in 1581 CE. By 1850 CE, it had fallen to 39.5 years and for some sub-groups, particularly those in the industrial cities, the decline was even greater. Death rates exceeded birth rates, driving down average life expectancy at birth as seen earlier for the years 1826 to 1850 CE. Friedrich Engels, economist, social critic, and co-author with Karl Marx of *Das Kapital*, described these conditions as "social murder."

In *An Inquiry into the Sanitary Conditions of the Labouring Population* (1842), Britain's leading mid-nineteenth-century reformer on issues of water quality and sanitation, Edwin Chadwick, wrote: "various forms of epidemic, endemic, and other disease [are] caused chiefly amongst the labouring classes by decomposing animal and vegetable substances, by damp and filth, and close and overcrowding dwellings [that] prevail amongst the population in every part of the kingdom, whether dwelling in separate houses, in rural villages, in small towns, in the larger towns – as they have been found to prevail in the lowest districts of the metropolis."[27]

Although the rural towns and villages of England experienced lower mortality rates than the national average and slightly lower than London, it was in the growing northern industrial cities, such as Leeds, Birmingham, Sheffield, and Manchester, that high mortality rates exacted a lethal toll on the working classes. Chadwick estimated that 13 years of productive life were lost due to the unsanitary conditions of these cities. Recent evidence, however, indicates that he underestimated the impact of urban industrial life in the early decades of the nineteenth century.

Since a defining characteristic of industrialization is the movement of people, gains in wages must have been the motivating consideration despite the increased cost of housing, food, and clothing. Hans-Joachim Voth in his analysis of the relationship of moving, wages, and mortality described it in the following way. "The average 25-year old man in England in 1841 CE had a 49.4 percent chance of living to the age of 65. By moving to Bristol (by no means the worst of the industrial cites) his chances would decline to 34.9 percent. A 15.5 percentage point higher chance of death by age 65 is the 'physical price' of moving – for every seven Englishman per cohort dying in rural areas, nine would be dead by age 65 in the cities. In the north, men in the cities had to last to age 56 to enjoy the same cumulative earnings of their rural peers between ages 25 to 65. Fewer than 40 percent of each cohort of 25 year olds did. In the south, they only had to live to age 43; approximately 64 percent managed."[28] In the case of Manchester rather than Bristol, a smaller percentage lived to benefit from the move to cities. In 1841 CE, the average life expectancy in Manchester was 26.6 years and in Liverpool 28.1 years. Since nineteenth-century migrants lacked perfect information about moving before making their choices, wage gains for many seem to have been canceled out by the conditions that awaited them.

National awareness of the lack of amenities resulted in some major improvements during the 1850s despite the continuation of overcrowding, overwhelmed public services, and a general lack of sanitation and hygiene. Paving main streets, adding gas lighting, naming streets, and identifying houses by numbers became more customary in British cities.

To reduce the risk of fires, brick and tile construction replaced buildings and dwellings made of wood with thatched roofs. Separating living zones from places of work reduced the noise, congestion, and industrial pollution on residents. Public buildings, including museums and libraries, began to appear as amenities for the wellbeing of urbanites.

British urban reformers, appalled by the congestion, noise, and smoke of factories, labeled them "dark satanic mills." They were responding to the outward realities of the physical environment and probably unaware of the declining rates of infant mortality that translated into declines overall. The smoke pouring forth from hundreds of residential chimneys symbolized cheap affordable coal for factory workers and their families. After the 1850s, households were warmer in winter than those of their rural brethren, they ate better-cooked food, and their cleanliness measured by eighteenth- and nineteenth-century standards was superior. Despite "environmental ugliness," as measured by the standards of the time, including the increased mortality of newcomers to the cities, Britain's surviving urban factory poor experienced improved living conditions.[29]

The Power of Steam

The transition to steam power and the burning of coal was a response to the introduction of new machinery in the workplace that required more and more power. The competitive demands to produce more and more goods quickly, efficiently, and at reduced labor costs accelerated technological innovation. First invented as machines to pump water out of flooded coal and metal mines, steam engines were the transformative machines that prompted the separation from early manufacturing and proto-industrialization to a factory-centered industrialization. They also raised water to operate waterwheels and for public water supplies. Despite the dominance of waterwheels, steam engines had become early symbols with mechanization by the 1770s.

Since industrialization was an evolving process, the replacement of waterwheels with steam engines was a slow process. In 1800 CE, waterwheels outnumbered steam by a ratio of four to one and for the next 50 years continued to increase their share of the energy market. Britain was a well-watered environment with currents flowing in its countless rivers and streams. Coal was plentiful and cheap but an imported commodity requiring transportation in an age before an extensive rail-road system. "Water wheels in sheer numbers and variety of uses remained the dominant power unit from 1755 to 1830."[30]

Agriculture remained the largest occupation in Britain in 1850 CE, followed by domestic service. The country employed more construction workers than cotton mill workers. Shoemakers outnumbered coal and metal miners, as did blacksmiths when compared to ironworkers. An industrial evolution was in the making, having achieved adolescence but not yet maturity. As late as 1870 CE, the heavy use of steam engines was centered in three industries: mining, textiles, and metallurgy. All others depended as much or more on waterwheels.

The last 30 years of the nineteenth century saw a tenfold increase in the use of steam power. The railroads provided an additional impetus with their use of steam boilers to power their locomotives, thereby allowing people and product to move at a speed of 50 miles per hour as opposed to the one to five miles per hour that existed previously. By the 1870s, most European countries possessed networks of railroads connecting the countryside to the city and city to city.

Using steam-powered pumps to eliminate water from coal mines allowed miners to sink deeper shafts to extract more coal. The cost benefits of digging deeper were apparent to mine operators. Coal dominated this phase of industrialization. The environmental effects of digging deeper in terms of soil degradation, land subsidence, and debris left on the land eluded early industrialists as they sought to minimize their costs and maximize their profits. Shafts between 1,000 and 2,000 feet became commonplace in the early years of the nineteenth century. Metallurgy, practiced for thousands of years before industrialization, experienced a transformation with the use of steam engines. They provided power for iron and steel blast furnaces, for mechanical hammers to stamp metal, and for rollers that flattened and shaped sheet metal.

The relationship between the extraction of coal and the production of iron was a symbiotic one. The deforestation of the countryside in order to produce the tons of charcoal needed to manufacture weapons, utensils, ornaments, and works of art has been documented at length in the last chapter. Once a fuel crisis hit, an alternative fuel, in this case fossil coal, would catapult Britain, Europe, and the United States into industrial world prominence. For example, in 1800 CE, Britain's per capita production of iron reached 26 pounds. Eighty years later, the figure grew to 260 pounds. Pig iron production increased eightfold in the last 50 years of the eighteenth century and by a factor of 30 during the nineteenth century. The mining of coal had to outpace this extraordinary growth in production. Fewer than three million tons of coal was mined in 1700 CE. By 1800 CE, the figure was 10 million tons and 50 years later it was over 50 million.[31] France and Germany experienced a similar growth spurt in the mining of coal. In 1870 CE, French miners extracted 13 million tons of coal compared to one million tons 30 years earlier. In Prussia (before the unification of Germany in 1871), coal production rose from 1.5 million tons to 20.5 million from 1825 to 1865 CE. Britain's coal output peaked in the years from 1910–14 at 270 million tons.[32]

Mining coal, the product of compressed plant and animal biomass and buried for as long as 550 million years, changed the landscape of the countryside. A German visitor to England vividly described the poignant image of coal in the countryside in 1844: "Imagine black roads winding through verdant fields, the long trains of wagons heavily laden with black treasures … Burning mounds of coal scattered over the plain, black pit mouths, and you will have a tolerable idea of what the English delight to call their Black Indies."[33]

The image, however bleak, portrayed a vibrant Manchester in 1814 CE. "The cloud of coal vapor may be observed from afar. The houses are blackened by it. The river, which flows through Manchester, is so filled with waste dyestuffs that it resembles a dyer's vat. The whole picture is a melancholic one. Nevertheless, everywhere one sees busy, happy, and well-nourished people, and this raises the observer's spirits."[34]

The impact on air quality would become transformative. Residential and industrial burning blackened the skies, killed vegetation of all kinds from trees to weedy plants, and stained the built environment with tar and soot. Although apologists for the smoke would represent it as a symbol of work, progress, and industrial might, knowledge about its emission of greenhouse gases – carbon, sulfur, nitrogen oxides – would await the invention of scientific instruments to measure their presence at the molecular level. In the past, many others would celebrate the medicinal benefits of smoke for any number of respiratory ailments, but much later in time, science would prove unmistakably that city dwellers suffer respiratory illnesses in greater numbers than those in the countryside as a result of toxic emissions.

The Role of Invention and Innovation

In Britain and in Europe, the growing role of fossil fuels to power steam engines helped to define the first century of industrialization from 1760 to 1850 CE. However often this statement is made, it fails to communicate the physical and intellectual energy required to distinguish this first century from all previous ones and to propel industry forward into the centuries that followed. Finding coal seams, digging, transporting it by cart and later by railroad to industries that converted the heat released in burning into kinetic energy to do work required physical strength and useful knowledge. The invention of the miner's lamp to reduce coal-mine fires and a more efficient stove also became part of a larger connection between useful knowledge to improve material well-being and the growing relationship between practical "things" and abstract scientific principles.

This useful knowledge came in chunk-sized bits, through practice, trial and error, and with an artisan's know-how by tinkering, and in building and rebuilding devices and apparatuses. For example, the physics of steam power was not known totally by its early inventors. They knew about atmosphere and atmospheric pressure, however, and they built an atmospheric device that converted heat into work. They also knew that the vacuum created by converting water into steam created the pressure that forced a piston to move in a cylinder.[35] Many other bits of knowledge were known incompletely but well enough to carry forth an inventive process that may have begun as early as the twelfth century with the invention of the spinning wheel and the horizontal loom. As noted earlier, in the fifteenth century the invention of the blast furnace reduced the amount of charcoal needed to produce iron.

What distinguished these innovations from those invented in the past was the momentum that they created. Past experience proved to be much more cyclical, with periods of progress followed by long periods in which wars, famines, and epidemic diseases retarded forward movement. The early inventions of industrialization were followed after 1820 CE with others that continued the early momentum. They included the adaptation of spinning, weaving, and carding of cotton to wool and linen. The uses of the hot blast purified and refined iron making. The building of bigger and more efficient steam engines for both stationary low-pressure engines and for high-pressure engines of locomotives advanced industry and transportation.

At the same time, the invention of hot-air balloons became an early version of flight; vaccines for smallpox began a long scientific and medical journey to conquer infectious diseases. Canning food can be thought of as an early form of processing without chemical additives. Ancient forms of manufacturing were replaced by new and sometimes more elegant designs and patterns.[36] Major inventions and numerous micro-inventions provided the impetus for a continuation of industrial change. The production of steel and the harnessing of electricity would propel industrialization into future centuries. But first, we should examine the impact of early industrialization and its environmental effects in the United States.

Comparing Industrialization in the United States and in Britain

As early as the 1830s, women in their New England homes were making buttons, hats, and shoes under contract to outside distributors. In European households, this system of working at home under contract represented a step on the road to a system in which the factory replaced the household. In the United States, both systems functioned simultaneously until industrialization in some areas replaced the other in the first half of the nineteenth century. For long periods, many different economic arrangements persisted as households, work outside of the family, and small manufacturing establishments functioned simultaneously as makers of boots, harnesses, and a diverse array of leather goods. They competed with their equivalents in glassblowing, metal smithing, pottery making, and spinning of textiles for consumer dollars.

As happened in much of the late eighteenth- and nineteenth-century world of industry, family commerce failed to keep up with the demands of a growing population that looked outside of the family structure for affordable and standardized consumer goods. Given the growth trajectory of the population, household economies and outwork arrangements simply couldn't meet a demand that large-scale industry would eventually satisfy.

In Europe and the United States, merchants accumulated capital in commercial trading activities including lumber, grain, molasses, sugar, tobacco, cotton, fish, and animal furs and pelts. All became part of a hemispheric economy that became global and gave new meaning to the term commodity, as natural products became goods for consumption. Turning humans into commodities, as happened with the spread of the Atlantic slave trade and the expansion of wage labor, reflected upon the expansiveness of this new terminology. Capital accumulation became a priority as entrepreneurs became investors in new untested industries. Today, we would refer to such individuals as venture capitalists. The farm Enclosure Acts in Britain drove laborers off the land into the cities. In the United States, growing households and declining farm size through generational bequests drove siblings to find employment elsewhere. With too many people in the rural households, migration provided new industries with an available and eager labor force.

In this transition to industry, hard work did not mean that artisans, servants, and free laborers shared equally in the benefits of their work. In Philadelphia, the country's largest city during the colonial era, income inequality was the norm. There, the top 10 percent based on wealth owned 90 percent of the taxable property, while 30 percent of the population lived in poverty.[37] Conditions in the plantation South revealed even greater disparities. Tobacco plantations replaced the declining pool of indentured servants with slaves. And the gap among the various social classes grew greater as large plantation owners squeezed out yeomen. In this way, they enlarged their holdings as the demand for cheap tobacco grew in Europe. In the process, the repeated planting and harvesting of a mono-crop destroyed the vitality of the land and the diversity inherent in crop rotation.

Despite the vast numbers of people invested in the land as freeholders, yeomen, and plantation masters, the numbers released from the land to seek employment elsewhere provided industry with its workers. Promoters of industry targeted women and children, since these boosters believed that young women and children represented surplus population with no visible means of support. As such, they accepted industrial work as a substitute for poverty and starvation. Benjamin Franklin recognized this tradeoff early by noting that, "it is the multitude of poor without land in a country, and who must work for others at low wages or starve, that enables undertakers to carry on manufacture."[38] Although Franklin and Jefferson were inventors and intrigued by machines, they believed that manufacturing was a divisive and corrosive element in a new country trying to avoid the pitfalls of the social inequalities they witnessed in Europe.

For them, industry was rapidly changing Britain into a nation of city dwellers. And although the transformation would not become apparent to many until the early decades of the nineteenth century, Jefferson and Franklin saw the trend early on. The early nineteenth-century transformation was quite remarkable. The invention of the steam boiler rapidly became the power plant for new steam-operated railroad engines and eventually steam power replaced water-driven wheels in the factories. Jefferson and Franklin saw the beginnings of these trends in Britain and wanted the new nation to avoid them. Their vision of an agrarian republic in the United States was one that would sustain the democratic values of thrift, self-sufficiency, and slow growth.

What Jefferson and Franklin could not foresee was that behind the grit and grime, so noxious to their senses, was invention, innovation, entrepreneurship, and economic ferment unknown in the rural economy. In such a place, learning accelerated as knowledge became the vehicle for progress. While all was in upheaval, few could predict what the future would hold but all had to know that a new society was unfolding – one that was volatile, unpredictable, one with winners and losers, and one in which fortunes were amassed and lost. Aristocracies didn't suddenly disappear in this new industrial world. Many, however, became members of the investor class providing the capital for industrial innovations; some aristocrats watched their inherited power and privilege erode as wealth based on achievement replaced inheritance as the road to success.

Pristine environments posed serious production problems for investors and manufacturers, however. Mills dependent on hydraulic energy curtailed production when rivers dried up in the dog days of summer and froze during long winters. Seasonal weather conditions, not the supply of raw wool or cotton or the demands for finished goods by a growing urban and rural population, stymied production and affected revenues and profits. Moving to environments where larger rivers drew from watersheds that covered hundreds and even thousands of miles, where dams could be constructed to hold back water, and where the topography produced waterfalls would result in increased production and lower costs. In the longer run, smaller village mills would lose out to larger facilities.

In some sense, these village mills mimicked the patterns of seasonal agricultural life. The number of daylight hours shaped the rhythm of work, sleep, and routine activities. Snow-covered fields drove farmers and village millers into the forests to cut and log timber and to replenish their woodsheds with fuel for cooking and heating. Spring and fall became times to plant and harvest, while the summer heat required cultivation, weeding, and watering. Industry on a larger scale changed human relationships, their relationships with the seasons, and the dynamic interaction that farmers and villagers had with the natural world.

The location of larger-scale machine industry in cities meant that household members became specialists in their work roles. The factory replaced the village cottage as the locus of work. Even though household members continued to produce goods for their own consumption, homespun and outwork disappeared as economic activities. The factory now became the primary workplace and family members became separated from each other during a typical day. Learning new work skills and using new technologies of production required a new and higher level of literacy. Skilled workers earned higher wages, ate better food, received medical attention for their ills and injuries, and lived longer. Unskilled workers received lower wages, ate poorer quality food, lived in more congested neighborhoods, could not afford medical attention for illnesses and injuries, and died sooner. Infectious disease, illness, and the imposition of longer and more intense hours of work led the skilled and unskilled to note that, "they never work so hard as when they entered the factory."

The impact of these changes, not initially apparent to those who experienced them, was an increase in infant mortality in the generations immediately following the rush to cities. In addition to higher rates of infant death, a decline in fertility occurred. Agriculture and village cottage industries thrived on large households with children engaged in work. Early machine industries were not only incubators for disease and sinks for airborne and waterborne contamination; they were also environments in which too many children placed heavy economic burdens on families. Birth control became a common practice.

In eighteenth- and nineteenth-century Europe, it took decades for industrial production to surpass agriculture. So, for many decades rural industry that produced for the household and the market coexisted with the urban machine industry that produced solely for the market. The transition took place earliest in Britain by 1810 CE but later in France (1840) and much later in Germany (1890).[39] In the United States, agricultural and industrial productivity remained on a par until the early twentieth century. As a result, the idea of an industrial revolution has dramatic appeal but it may be wiser to think about the process as an evolutionary one.

As noted earlier, there was no single path to the development of industry in the United States in the nineteenth century. Nor was it limited to one region of the country. Other forms of industrialization existed alongside New England's textile behemoths. Manufacturing clothing, metals, paper, boots, shoes, and machine tools took place in the metropolitan mills of New York City, Philadelphia, and Baltimore. Many lesser rural towns along the Connecticut Valley and small cities, such as Springfield, Chicopee, Pepperell, Massachusetts, and Saco, Maine, joined them. In all, there existed many models of industrialization, from light handicraft to hydraulic power mills of many shapes and sizes in rural and urban America.[40]

The one path whose history is preserved in a national park setting is the one-industry textile town exemplified by Lowell, Massachusetts. For most Americans, this one-industry town symbolizes the age of industry almost 200 years ago. It built on the successes of an earlier and smaller mill town, Waltham, Massachusetts, in 1814 CE that harnessed the waters of the Charles River to power two mills. Its principal owners, Francis Cabot Lowell and Nathan Appleton, employed young, unmarried women from established farm families to work as weavers and spinners. The natural flow of the Charles River could not satisfy the demands of these two textile mills for waterpower, however.

The search for a more powerful river with a greater annual flow and able to power many more mills commenced. And with this search began the dependence of the United States on water to power its industries into the middle of the nineteenth century, despite the invention of the steam engine much earlier. Even with the emergence of fossil fuels in the twentieth century, 20 percent of American industries depended on waterpower in 1900.[41] The owners of the Waltham mills found the watercourse they needed on the Merrimack River, a river that the environmental historian Ted Steinberg identified as "perhaps the most celebrated river valley in America's early industrial history."[42]

The valley includes 5,000 square miles of land and water. The river travels 116 miles and on the lower part the average daily flow reaches 6,000 cubic feet per second.[43] Within the first 50 years of the nineteenth century, the cities of Lowell and Lawrence in Massachusetts and

Manchester and Concord in New Hampshire would become the country's industrial centers for the production of textiles.

To accomplish the goal of industrial dominance in textiles, this wild and scenic river that ran unobstructed to the ocean had to be controlled by constructing dams, locks, and diversion canals that moved the river's waters into the basements of the mills to move the waterwheels that powered machines for carding, spinning, and weaving cloth and rugs. This massive effort to control and harness the river's energy symbolized progress in the minds of mill owners. Not only did they produce sturdy cotton and woolens for a growing national and international market, they provided employment for a surplus population of unmarried women and children.

Mastery over nature symbolized by the rushing river required human intervention and ingenuity. "The natural world existed as a reservoir of productive potential awaiting the contriving hand of humanity."[44] By 1850 CE, over 40 mills operated in the valley and produced more than 14 million dollars in cotton textiles. Where hundreds were employed when controlling the river began in 1821 CE, thousands were now employed. In the new nation, between 1820 and 1840 CE, the number of people employed in industry doubled and investment in manufacturing reached almost 40 percent of the investment in agriculture.[45]

For visitors to and observers of this new industrializing America, even those who voiced reservations about the transformation from agriculture to industry, their expressions were those of promise and aspiration with mills springing out of the ground. As the Swedish novelist Fredrika Bremer approached Lowell on a winter's night she described the city in idyllic terms: "I had a glorious view from the top of Drewcroft Hill, in that starlight the manufactories of Lowell glittered with a thousand lights. They were symbols of a healthful and hopeful life in the persons whose labors they lighted; to know that within every heart in this palace of labor burned a bright little light, illuminating a future of comfort and prosperity which every day and every turn of the wheel only brought the nearer."[46]

Witnesses to the coming of the factory also celebrated the impact of industry on its workers in equally positive language. As agriculture was represented for its cleansing and empowering qualities, so manufacturing received similar praise. As the social commentator Harriet Martineau noted, "I found in some places very bad morals had prevailed before the introduction of manufactures; while now the same society is eminently orderly."[47] Similarly, the world traveler James Silk Buckingham commented that "the health, both of males and females, but especially the latter, is much better after being employed in the factories here than before entering them, though they almost all come from the farming districts of the surrounding country."[48] Although evidence collected later about the health of women workers in the mills would contradict these assertions, the initial and immediate impression prevailed – the factory towns were new, clean, and healthy.

Nature, as defined by the river's extensive watershed fed by a northern lake system and a region with ample annual rainfall of 45 inches, could not withstand the assault by industry. The river's natural monitors of its health are the aquatic and plant life that thrive in healthy waters and die off as rivers became contaminated. By 1867 CE, the New England Commissioners of River Fisheries reported that, "a half century ago, [salmon, shad, alewives, and others] furnished abundant and wholesome food to the people; but by the erection of impassable dams, and needless pollution of ponds and rivers, and by reckless fishing … our streams and lakes have been pretty much depopulated."[49] As more and more dams were constructed in the twentieth century, dams impeded the return of fish to their spawning grounds and ultimately led to their demise. The construction of the massive 30-foot-high Lawrence Dam in 1848 CE became the major barrier to a healthy Merrimack.

The Merrimack became a sink for industrial wastes, as textile mills and tanneries released dyes and chemicals, lumber mills washed away their sawdust, and humans flushed their accumulated waste into the river. In 1870 CE, Lowell's population passed 40,000, Lawrence approached 29,000, and Manchester, New Hampshire, exceeded 23,000.[50] As the cities along the river continued their economic advance, the river's ability to purify itself by receiving clean water from its upstream watershed declined. Those who lived through this transformation recognized the causes and the effects. From Lawrence, an observer wrote in 1890, "Ten years ago the water of the Merrimac River was much purer than it is today. Every year sees more refuse in the river from cities above us, and the water is, of course, more polluted all the more … the poisonous dyes and mill refuse also find an outlet in the river. This disease-laden water flows only nine miles, and then the people of Lawrence drink it. No wonder that disease and death follow."[51]

The transition to steam-powered factories signified the decline in waterpower. Mills along the Merrimack changed to keep up with the newer textile mills using this power source. They lost their comparative advantage, however, to those who began from scratch, not those like themselves who sought to renew and retrofit themselves to industry's latest technological innovation. Industrialization was about progress as its investors and principals saw it, not about conserving and adapting. The hydraulic energy of the Merrimack River had powered industries that made cloth, rugs, iron tools, leather, lumber and wood products, and many more, for more than half a century. Now, steam power and the dependence on fossil coal would power the next wave of industry. According to the French topographical engineer Guillaume Tell Poussin, "Steam, with the Americans, is an eminently national element, adapted to their character, their manners, their habits, and their necessities."[52]

Manufacturers had taken pristine nature and put it to human use. That use created great fortunes for owners and investors, wages for thousands of workers, and economic benefits for both. Cities were constructed where wilderness existed and they became living spaces as well as centers of culture and learning. Despite the hardship of industrial life, many benefited from the new social and economic arrangements.

One-industry mill towns continued to dominate the landscape. The textile mills of Lowell continued to operate but they lived now in the shadows of the newer and massive woolen mills of Lawrence, where three of the country's four largest mills were located. By 1899 CE, all would be combined into a single corporate entity, the American Woolen Company, employing 12,000 workers. North of Lowell, the Amoskeag Manufacturing Company in Manchester, New Hampshire, became the world's largest textile mill, comprising 30 buildings and 17,000 workers. To the south of Lowell, Fall River, Massachusetts, emerged as a textile powerhouse, despite the absence of a fast-flowing river. Using steam-powered and automated machinery fed by coal from Pennsylvania, Fall River became a prototype for textile mills freed from the hydraulic energy of fast-moving rivers in the United States.[53] The watershed suffered, however; its tributaries became polluted, disease spread, and the river's ability to renew itself declined. In the short run, industry blossomed in New England and throughout the Middle Atlantic States and Midwest. In the long run, the Merrimack Valley's economic miracle would become a nightmare as the region spun downward into a long-term economic depression.

Figure 7.1   Young cotton thread spinners in Georgia, 1909. *Source*: Lewis
W. Hine (LOC). http://www.flickr.com/photos/39735679@N00/484535150/.

The enthusiasm expressed by European boosters about the pure and
regenerative qualities of anti-bellum factory work in Lowell and elsewhere seems overstated when examined in light of the evidence gathered in the post Civil War period. The factories that they visited in the
1840s and 1850s paled in comparison to the consolidated works and
new mega-factories built after the War. This change in magnitude may
help to explain the grueling impact of industrial millwork on the labor
force. The evidence gathered for the period 1905–12, the first of its kind,
suggested the following. Workers in New England's cotton textile mills
died at rates higher than non-mill workers. Longer years worked in the
mills increased rates of mortality. Persons with fewer occupational
choices, namely immigrants, young men and women with limited education, suffered the highest death rates. The combination of millwork
and homemaking overwhelmed the wellbeing of women mill workers.
Death rates were the highest among married women who gave birth to
children while working in cotton mills.[54]

The idealized cotton mill of European visitors disappeared when
exposed to the realities of grueling work that resulted in fatigue and a

lowered resistance to the common cold, making workers vulnerable to tuberculosis and finally death. With a normal workweek of 54 to 58 hours, working at very close quarters in a very hot and humid environment, the factory became an incubator for the spread of infectious disease. To maintain the necessary high humidity and reduce friction, a jet of steam was vented directly into the weaving rooms where workers tended the looms. Workers not only drank water from a common cup, weaving room attendants "kissed the weft" whenever a shuttle was empty. To replenish the shuttle, thread would be placed next to the eye and literally sucked through by a weaver's well-placed lips. Since many different weavers attended many and the same looms, "kissing the weft" became a most efficient method for spreading infectious diseases. The invention of the automated loom eliminated this method of threading but it increased the workload of weavers, replacing "kissing" with additional fatigue and stress in the mill room.[55]

The centralization of textile mills into vehicles for the mass production of cotton cloth, wool, and linen changed the metabolism of cities that housed them. Population density rose significantly. Propinquity translated into overcrowded housing, congested streets, and factories teeming with human and mechanical activity. These densely populated centers for industry became breeding grounds for endemic disease. Outbreaks of cholera and typhoid fever caused by feces-laden water regularly visited nineteenth-century cities. The common cold, influenza, pneumonia, the pox in all its forms, scarlet fever, whooping cough, mumps, and many others were a constant reminder of the lack of sanitation and health in cities. Add to these conditions the vulnerabilities accentuated by work in the mills and we have human suffering and premature death. In brief, we have human tragedy.

Beyond the mill towns of New England, the economy of the Atlantic coast was booming with activity. Wilmington, Delaware, turned animal hides into consumer goods and became a major manufacturing center for tanning morocco leather. Using the moving waters of the Delaware River as a sink for its tanning wastes, it gained international prominence for its products. Trenton, New Jersey's Iron Works became a leader in the production of steel beams and cable in the construction of buildings and suspension bridges including the Brooklyn Bridge. Iron facades for commercial buildings and steel beams for tall buildings transformed the built environment. Its cable bridges added to the mix of modern transportation systems.

Since the Trenton area contained ample amounts of silicate minerals found in igneous and metamorphic rocks and clay, it became a center for the production of ceramics. With the nation becoming increasingly aware of the need for improved sanitation in the early years of the twentieth century, sinks, tubs, and toilets became principal products of the city's ceramics industry. By 1900, the city employed 15,000 workers in 40 pottery works.[56] Cities that possessed readily available natural and human resources thrived during periods of economic growth. If they had access to any of the following resources or a combination of them, then they possessed the potential for success: a moving river for hydraulic energy and a sink for wastes, an adaptable workforce drawn from surrounding villages and towns or from distant places, woodlands for charcoal, building materials, and for fuel, and animals for their hides and protein. In the case of Trenton, a soil with a geological history and the chemical composition ready to be crafted by human hands into products to improve the human condition gave the city a comparative advantage in the industrial marketplace.

In Europe and the United States, coal production played a central role in the drive to establish concentrated, large-scale industrial enterprises in the second half of the nineteenth century. Governments and financial markets, including banks, helped to create national railroad networks powered by the energy released by coal combustion. By 1875 CE, large steel mills "burst from the ground, like mushrooms, in the valleys of the Rhine and the Ohio Rivers and their tributaries: the Ruhr, the Saar, the Moselle, the Allegheny, the Monongahela, and the Mohoning."[57] Their tributaries and rail systems carried coal, coke, iron ore, limestone, and steel products, including machinery, to other industrial districts. The markets for their products spread over wide areas, crossing national and international boundaries. In addition to the larger nations of Europe, the smaller ones, including Switzerland, Sweden, Belgium, and the Netherlands, each with different natural resources but with access to common technologies and transportation, experienced a 15 to 25 percent increase in national income each decade from 1850 to 1900 CE.

Coal, Iron, and Steel

As the world's leading workshop, Britain led the world in coal production for many centuries. And with dwindling supplies of timber as a fuel source, coal came to the rescue and powered Britain's continuing industrialization. During the late nineteenth century, coal also powered the accelerating industrial might of the United States to become the most visible source of urban industrial and residential pollution. Today, the country burns more than one and one-half billion tons of coal annually to generate electricity for consumers nationwide, outstripping all countries.

As Barbara Freese noted in *Coal: A Human History*, "Like the good genie, coal has granted many of our wishes, enriching most of us in developed nations beyond our wildest pre-industrial dreams. But also like a genie, coal has an unpredictable and threatening side. And, although we've always known that, we are just beginning to realize how far-reaching that dark side is."[58] For the darker side, emissions from coal-fired plants cause an estimated 10,000 deaths annually in the United States due to respiratory failure and lung cancer. As the world's largest producer and consumer of coal, the United States has passed nineteenth-century Britain's capacity many times over. In the early years of this century, China's insatiable energy demands will eventually make it the global leader in coal consumption, with similar health effects on its population.

The availability of natural resources often triggered the construction of factories, mills, and eventually complex industrial enterprises that stretched for miles along a watercourse. Although the development of an integrated railroad system in the late nineteenth century allowed for greater distances between resources and finished products, closeness remained an important variable in calculating an industry's competitive advantage. The mid-Atlantic states of Pennsylvania and West Virginia possessed some of the nation's richest deep beds of bituminous coal. Pittsburgh's claim to the title of "workshop of the world," producing steel for domestic and international use, depended on its magnificent coal resources. Without it, the city's industrial promise faded. As Willard Glazier wrote in 1883, Pittsburgh's iron and steel industry was "rendered possible by the coal which abounds in measureless quantities in the immediate neighborhood of the city."[59]

The demand for coal became insatiable as Pittsburgh's steam-powered railroads, factories, and industries expanded their manufacturing output. Four southwestern counties of Pennsylvania, the location of Pittsburgh's deep seam of coal, expanded production to meet the city's production needs. In 1850, they produced less than a half million tons of coal annually but by 1911 production had skyrocketed to 66.5 million tons.[60] Much of the coal was shipped directly to consumers, mostly to railroad and commercial enterprises and to households. Coal for iron and steel mills was another matter.

Carbon is required in the smelting process to produce iron. For centuries, iron foundries were located close to woodlands since the burning of wood produced the charcoal needed in the production process. With the depletion of forests and the discovery of coal, coking coal replaced charcoal in manufacturing iron and steel. With the expansion of rail transportation and the decline in charcoal production, blast furnaces, formerly located on the periphery of cities, were now located in the city. The production of coke, however, remained in close proximity to the coal mines and for about 70 years from 1850 to 1920 was produced in beehive ovens.

Understanding coke production in beehive ovens is important for a number of reasons, the most important being the effect of this production process on the surrounding physical environment. Similar to beehives, they were dome shaped. With a base diameter of 12 feet and about 7 feet tall, these brick enclosures with holes in the top allowed particulate matter and gases to escape under high temperatures. Cooking coal for two to four days, the high-carbon coke was removed from the ovens manually until about 1900 and mechanically thereafter. Hundreds of these ovens, arranged in rows, rose above the natural contours of the land and contaminated the air with plumes of noxious gases and tar-laced smoke. As one observer noted, "The spring smoke is apt to be whitish-blue, that of winter brown or gray, and any time of year it ranges from pinkish-lavender to gun-metal tinged with purple at evening."[61]

By 1909, as many as 104,000 beehive ovens, with over half of them located in southwestern Pennsylvania, were producing millions of tons of coke for the country's steel industry, also centered in Pittsburgh and the surrounding municipalities. Although many of the ovens continued to function in the countryside, the Jones and Laughlin Iron and Steel Company, located in the city, operated the world's largest beehive coking facility there with 1,510 ovens.[62] Locating the ovens near the blast furnaces undoubtedly cut production costs. The devastating cost to the environment was visible for all to see.

For miles around, beehive oven emissions covered the land with a blanket of dust, tar, ash, and oil. A botanist noted in 1900 that "the most conspicuous feature in the coke oven surroundings is the general wretchedness of everything of the nature of shrub or tree, either individual or collective."[63] Vented emissions caused much of the widespread damage but dumping wastes on the site of the ovens was equally destructive to the land and waterways. Solid coal wastes were dumped on the ground and into nearby streams and creeks, altering their flow, causing flooding, destroying aquatic life and the water's ecology. While the focus here has been on the production of coke for industry, coal production and combustion created their own hazardous environmental effects.

In the last 50 years of the nineteenth century, coal consumption doubled every decade until the United States surpassed Britain as the world's largest producer of coal. In 1900, coal provided 71 percent of the nation's energy while wood had dropped to 21 percent.[64] By 1940, a number of the nation's largest cities, including Chicago, Cleveland, Milwaukee, Pittsburgh, and St. Louis, gained between 85 and 92 percent of their energy from coal combustion. It provided the energy to move raw materials, including coal, and people by rail and boat, to manufacture finished commercial products, including steel beams, angles, plates and suspension cables, and to produce an ever-growing number of consumer goods, including the newly invented labor-saving appliances. In this regard, the nation's capital stock of railroads, steel-framed buildings, manufacturing plants, and ocean-going vessels increased significantly while the availability of consumer goods for the growing middle and working classes became affordable. The tradeoffs in terms of increased consumption and environmental stress were seldom in balance.

The pre-industrial rhetoric that coal smoke killed malaria, protected one's eyesight, and fended off all things inimical to human health disappeared once urban public health officers and social scientists surveyed members of the urban working class. Tuberculosis, pneumonia, and bronchitis were among the major causes of death in late nineteenth-century cities, not only in the industrialized United States but also in Britain and Germany. Although the poor were more vulnerable, pulmonary diseases crossed class boundaries in areas contaminated with smoke pollution. No one, not even the wealthiest, escaped the ever-present plumes of smoke discharged by locomotives, steamboats, factory smokestacks, and residential chimneys.

Mining coal for industrial and residential use had pervasive effects on the physical environment. Aside from the solid debris of slag and

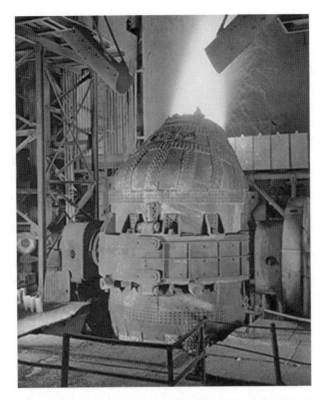

Figure 7.2　Bessemer Converter at Workington Iron & Steel located in West Cumbria, near the Lake District in northwest England. A blast of air through the molten pig iron in the converter removes impurities in the steel making process. *Source*: Permission granted by Phil Baggley.

rocks left from mining, sulfur acidic water flowing from the mines poisoned the rivers and streams, made the "hard" water undrinkable, and acted as a corrosive to all forms of piping and machinery that required either water to create steam or water as a coolant. For two of Pittsburgh's larger railroads, the annual cost for repairs, replacement, and cleaning amounted to nearly $800,000.[65] Residents substituted water from unlined wells for drinking and often exposed themselves to waterborne diseases. Outbreaks of typhoid fever were frequently attributed to citizens avoiding drinking treated tap water. Household plumbing also fell victim to sulfuric water, as did washed clothing.

Throughout the early decades of the twentieth century, local citizens, municipalities, and state governments fought to limit the most flagrant environmental abuses by the mining industry, the railroads, steam-powered barges, and factories. At almost every turn, their industrial influence and power exceeded that of government. Legislative action was weakened by powerful vested interests before bills became laws. Laws required enforcement by executive authority and protection from the courts. In the decades before the passage of the federal Clean Air and Clean Water Acts in the 1970s, municipalities and many of the states lacked the power to contain the abuses by many industries engaged in extraction, refining, and producing goods. Since federal laws require reauthorization by the Congress and enforcement by the President, legal protection required an alert and informed citizenry.

The clusters of nineteenth-century coke towns in the United States and in northwestern Europe were configurations that lasted into the twentieth century. They spread to Japan in the early decades of the twentieth century and to the former Soviet Union in the 1930s and their satellite countries in eastern Europe in the 1950s, where coke towns already existed in regions such as Silesia.[66] They existed to serve those huge steel plants that burst from the ground like mushrooms. Pittsburgh in the late nineteenth century was the center of the nation's iron and steel industry with more than half of its workers employed in these industries. By 1870 CE, the first vertically integrated steel mills were built there and for the first 70 years of the twentieth century, metal manufacturing accounted for 70 to 75 percent of Pittsburgh's industrial output. Although its share of the national steel market began to decline in the post World War II period, it remained a preeminent industrial city until the worldwide recession in the 1970s. By that time, steel-making overcapacity in the United States and intense competition from Japan and South Korea overwhelmed Pittsburgh's aging nineteenth-century mills.[67]

In the twenty-first century, China and more specifically its industrial and commercial cities of Shanghai, Beijing, Guangzhou, Shenzhen, Tianjin, Wuhan, Shenyang, Chongqing, and many others may claim the title of "workshop of the world." As recipients of the open economy initiatives of the central government since 1992, domestic and foreign investments have stimulated industrial development in iron and steel, construction materials, fertilizers, aeronautics, rail and auto transportation, consumer electronics, and apparel. More than a decade of assertive industrial development has lifted millions of Chinese citizens out of poverty and made them members of a growing global middle class.

Production of durable capital goods and the consumption of disposable material goods have left a distinctive ecological footprint on the land. Air pollution in twenty-first-century Chinese cities is reminiscent of factory towns in nineteenth- and twentieth-century Europe and the United States. Its contaminated air and that of its industrializing neighbors in India, Vietnam, and South Korea can be seen across the expansive Indian Ocean region. During the worst outbreaks, ocean-going vessels note its existence and take precautionary steps to avoid collisions. The rapid increase in automobile manufacturing and personal ownership adds significantly to declining air quality in South Asia. After exploring the emergence of auto mobility in the West, its global expansion will be described more fully.

Industrial Transformation and Global Auto Mobility

Motown (motor town) replaced coke town as the locus of American industry in the post World War II period. Titan cities such as Pittsburgh continued to produce steel for industrial and commercial purposes, with one of its principal consumers becoming the automobile industry. Automobiles replaced railroads as a form of motorized transportation and long-haul trucks competed with them for moving freight. Railroads had experienced almost 75 of uninterrupted growth but, in 1920, they stopped growing. With the rise of automotive transportation, the railroads began a decline that continued throughout the twentieth century. Before they replaced the electric streetcars in cities, automobiles had to displace urban horses as a form of transportation and as a beast of burden. In the United States in 1900, there was one horse for every four persons; in Britain, the ratio was one horse for every ten persons.[68]

In the nineteenth-century "organic city," horses became an integral part of the economy. They moved large quantities of raw materials, food, and manufactured goods within and around the surrounding countryside. This symbiotic relationship between city and country, town and farm integrated them into a unity unknown in the present. The relationship between animals and humans, commonplace on farms, continued when horses inhabited the same urban space. Their manure, a sorry sight on urban streets, was a source of noxious odors, "horse flies," and disease. On the streets of many large cities, at least 10,000 horse carcasses had to be removed annually. Their manure was collected daily and sold to truck farmers beyond the city's perimeter who fertilized their farms that produced oats for these same horses and vegetables for the masses crowding into urban America. Today, domesticated pets have replaced urban horses.

America's "love affair" with the automobile began slowly at the turn of the last century. Initially, it was seen as a noisy, unpredictable, and dangerous contraption. In contrast to defecating and urinating horses, its appeal became obvious to a generation of urban planners fixated on sanitation and public health. Although the pollution from millions of automobiles may be mostly invisible, its toxicity has led to emission standards in the last decades of the twentieth century and a ban on lead as an additive to gasoline. Also, as a major consumer of energy and manufactured materials in its construction, the automobile is an environmental disaster. "In Germany in the 1990s, the process generated about 29 tons of waste for every ton of car. Making a car emitted as much air pollution as did driving a car for 10 years."[69] Manufactured mostly of steel, iron, aluminum, and plastics, the industry consumed as much as 25 percent of the nation's productive capacity. More than half of the world's rubber plantations served the original and replacement tire needs of automobiles and trucks until synthetic rubber using petroleum in its production process replaced natural rubber during World War II. Trash dumps with pyramids of used tires are breeding grounds for mosquitoes and a constant fire hazard. New technologies for shredding tires for other uses hold some promise but the piles continue to grow and the hazards remain.

The automobile industry not only changed air quality and challenged the status of "King Coal" in the economy with fossil oil; it changed our relationship with the landscape. Millions of miles of roadways were constructed to accommodate this new technology with the building of the interstate highway system, beginning in the 1950s, to handle the flood of vehicles purchased by newly affluent Americans. Concrete and asphalt producers flourished. The extraction industry became a major beneficiary as limestone, sand, and crushed stone became necessary ingredients in producing composite materials. Coal tar, sand, stone, and oil combined to produce asphalt.

Construction trades changed the land's topography by employing thousands of workers using earth-moving equipment that grew larger and more powerful with each passing year. New roadways altered the

Figure 7.3 Hyundai Auto Assembly Plant in Korea where robotic technology has replaced human labor. *Source*: From Wikimedia Commons, the free media repository.

flora and changed the migratory patterns of land mammals. Land speculators and developers transformed abandoned farmland into suburban subdivisions. Horizontal development dispersed a growing population across a broad landscape.

The vertical city in the Northeast, the Mid-Atlantic states, and the Midwest began to empty as suburban sprawl and the concept of "land development" infused open spaces with the need to build housing sub-divisions, shopping-malls, secondary roads, parking lots, garages, and the ubiquitous restaurant chains serving the same fare no matter where you lived in America. Cars took lots of space, between 5 and 10 percent of the usable land, and altered the way in which drivers thought about the land. By 2000, the world had passed one-half billion in automobile ownership.

Worldwide, the automobile industry became the principal consumer of steel. By 2000, the world was producing more than 55 million new

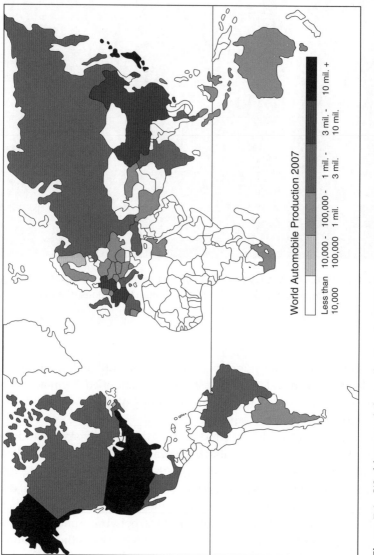

Figure 7.4 World automobile production for 2007. *Source*: Data from Organisation Internationale des Constructeurs d'Automobile. http://oica.net/category/production-statistics/.

vehicles each year. Japan eclipsed the United States in annual production in 1980 and continues to capture increasing market share with each passing year. In 2000, Japan produced 20 million automobiles followed by the 13 million produced in the United States. Germany was a distant third with fewer than 5 million vehicles, followed by France, South Korea, Russia, Italy, Canada, Spain, Britain, and Brazil. In the first decade of the twenty-first century, China and India with their populations in the billions and high economic growth rates will become producers and consumers of cars. Without new efficiencies, these internal combustion vehicles will add greatly to the global carbon load and accelerate the emission of greenhouse gases into the atmosphere.

CHAPTER EIGHT

TRADE AND CONSUMPTION

Introduction

"On a cozy summer morning in London, a successful merchant arises to meet the day by pulling off a chintz quilt, adjusting his muslin nightshirt, and covering himself with a Chinese silk dressing gown. At the appropriate time, a maid enters his room carrying a coordinated blue and white china tea service that holds tea, milk, and sugar. Unknown to the merchant, his country cousins had already purchased much finer imported patterns of calico and chintz not yet available in London."[1] All of these material goods became highly visible consumer purchases during the eighteenth century in Europe, a century referred to by some historians as the beginning of the consumer revolution.[2] Sugar, tobacco, tea, cocoa, coffee, porcelain, and Indian and Chinese textiles (silks, chintzes, muslins, calicos, and many others) became part of a growing market for consumer products in Europe stimulated by affordable prices, the desire for exotic goods, and the glamorizing of consumption among a growing urban population.

Consuming salable goods and merchandise became significant economic and cultural transactions with environmental implications beginning in the eighteenth century. Markets for goods and merchandise have a much longer history, however. With major silver mines in Iberia, the Romans exported coin and bullion and imported Chinese silks, pepper, cinnamon, cloves, and nutmeg. This was not the only way of paying for Asian goods, however; Roman merchants exchanged fine glassware, luxury ceramics, metal goods, textiles, wines and oils, and slaves in the Indian Ocean trade.

Global Trading Networks

Independent of trade between West and East, a vibrant trade among Indian Ocean civilizations provided spices, silks, cotton textiles, porcelain, glass, and precious stones to regional markets in eastern Africa, the Indian subcontinent, China, Indonesia, and Japan. Overland and maritime exchanges of raw materials such as dyes, skins, and hides as well as basic metals – iron, tin, lead, and copper – promoted thriving economies throughout the Indian Ocean, many centuries before the arrival of competitors from Portugal, the Netherlands, and Britain. As one historian has noted, "Variations in consumer tastes actually increased the flow of inter-regional trade and provided merchants with opportunities for high profits. Gastronomic traditions as they developed in the Indian subcontinent, Persia, and the Arab world held in high esteem the three finer spices of the Moluccas, cloves, nutmeg, and mace, and added the cinnamon of Ceylon to the list. In all these areas, rice dishes prepared with some of the spices were considered indispensable to aristocratic tables and festive occasions."[3]

In Europe centuries later, Italian merchants supplemented their ample supplies of silver with gold mined in sub-Saharan West Africa and purchased from North African Muslims in Cairo, Egypt. This major trading port connected Africa, Asia, and Europe with the great caravans of West Africa's kingdom of Ghana and its Mali Empire. The fruits of Africa's productive agriculture, its gold mines, and its thriving ironworks and slave trade entered the global economy. For 400 years, from the tenth to the fourteenth centuries, merchants and traders linked the continents together.

By 1252 CE the city-state of Florence minted the gold florin and Genoa did the same with the genovino. Thirty years later, the gold ducat from Venice appeared. All weighed about 3.5 grams and competed with other currencies throughout the Mediterranean world and beyond. Much of the history of this emerging global economy in consumer goods reached a highpoint when Marco Polo, along with his father Maffeo and uncle Niccolo, set off from Venice in 1271 CE to visit the Mongol Emperor, Khubilai Khan, at his summer palace, Shàngdu, on the Mongolian steppe.

As ambitious Venetian traders, the Polo family members hoped to return from China with silks, jewels, and much more for the market-places of this Mediterranean city that linked the West to the East and to bring back information about future trading opportunities. In fact, Marco Polo's *Travels* described commerce through Islamic Turkey, Persia, Afghan Central Asia, and the Silk Road cities along the edge of the Gobi Desert. It also detailed the potential rewards of trading with China proper. As an original travel guide, it provided advice to travelers and merchants about the difficulties of long-distance travel, goods available on the caravans along the way, provisions to stock, and the potential dangers presented by robbers.

Upon returning to Venice, the jewels sewn into their clothing confirmed what other merchants seeking fortunes suspected. And in the centuries that followed, Venice became the "liquid frontier" in trade between the Islamic East and the Christian West, often ignoring Roman papal bans on trade with non-believers. So many fortunes in trade were made that a Venetian diplomat acknowledged the city's symbiotic relationship with Muslims. "Being merchants, we cannot live without them."[4] A trading partnership brought spices, especially white and black pepper, and luxuries from India, China, and the Far East to the growing cities of northern and western Europe. In return, Venetians sold gold, silver, tin, and lead as well as furs, linen, and even hats in the East. Venice's power waned during the fourteenth century, as the ascendancy of the Ottoman Empire dominated trade in the Mediterranean and beyond. With it came the export to Europe of two new valuable and quickly sought-after products that challenged spices for the attention of consumers: sugar and coffee.

These earlier global trading relationships that created wealth for the few and brought luxuries to the elite members of Medieval and Renaissance society became the template upon which seventeenth- and eighteenth-century consumerism would expand and thrive. In Europe, the emerging middle classes glorified consumption but it was not limited to them, as masses of rural householders entered the cash economy of finished goods as well. As many historians have pointed out, consuming was truly an evolutionary development, much like that of agriculture, manufacturing, and industry described in earlier chapters.[5]

Distancing Consumers from Producers

Trading across great distances had begun in ancient times as consumers became uncoupled for the first time on a much larger scale from the natural world. The emergence of southern Mesopotamian cities in 8000 BP led to the growth of trading routes to resource-rich regions on the periphery. Timber, stone, oils, metals, and exotic goods entered these new cities and helped to buttress an increasingly stratified society. Wars of conquest were waged to gain control over neighboring populations and their resources, thereby extending Mesopotamian imperial control.[6]

Similar patterns of conquest and control of another's raw materials have become identified throughout history as city-states, nations, and empires engaged in warfare. Not all were fought with weapons, as the following narrative will point out. Some used aggressive forms of long-distance manipulation, including international capital, to gain control of a region's natural resources, engaging in "ecological imperialism." As such, a physical gulf began to develop, separating consumers from the natural world of production. Many assumed that nineteenth- and twentieth-century mass production and mass prosperity, symbolized by the ubiquity of the automobile, erased the linkages that formerly existed between economic exchange and the natural world. As we now know, the erosion of these natural connections between producers and consumers began much earlier. Examples of these uncoupling processes and their implications for the relationship between humans and the natural world will be discussed later in this chapter.[7]

People produced less and less of what they consumed in the century before industrialization and in the centuries that followed. They worked for wages and bought what they consumed. As capitalism flourished, individual wages and purchasing on credit expanded until the present era in which the mass consumption of food and material goods has eclipsed all benchmarks established in previous centuries. Here, a distinction between the mass consumption of food and material goods will become instructive for the following reasons.

The separation of producers of food from consumers began centuries ago, in the period 1460 to 1600 CE when long-distance trade in foodstuffs, particularly grain, animal flesh, and preserved fish, represented the transfer of harvested biomass and energy from one regional ecosystem to another. Ecosystem transformation created unequal gains and losses for the people caught up in this exchange. In a past era of local production and consumption, the concept of an "ecological footprint" had readily observable consequences. With the coming of long-distance trade in foodstuffs in particular, consumers became detached from the reality of the life and death of the biological organisms that they consumed.

Population growth and increasing urbanization created a greater demand for grains by 1450 CE. Clearing forests to meet the increased local demand for cereals also provided the opportunity to sell any surplus to distant buyers. Eastern European farmers sold their surplus grain, mostly rye cereals, during the sixteenth century to western urbanizing markets. Per capita meat consumption in the fifteenth and early sixteenth centuries reached a zenith that it would not achieve again until the late nineteenth and twentieth centuries. Annual per capita rates of meat consumption ranging from 120 lb. in Hamburg, Germany to less than 50 in Mediterranean towns were not uncommon.[8] Long-distance cattle drives became common during the late medieval and early modern periods of European history. A trend to supplement local meat supplies with those from the cattle production belt that extended from Europe's northern and eastern frontiers developed earlier and continued into the modern era.

Meat consumption continued to rise in Europe's western cities after 1500 CE as supplies from the eastern and northern frontiers increased. The long-distance cattle drives to foreign markets transferred nutrients from local pastures to roads and open fields along the routes to western cities. The removal of cattle and the hay and turf that they needed in order to maintain body weight and the loss of manure caused the loss of local soil fertility, bog formation, and created environments in which drifting sand replaced fertile soils. An ecological crisis resulted in cattle-exporting regions such as Denmark. By the late medieval period in sixteenth-century Europe, a pattern that would become recognizable into the twenty-first century took place. Hundreds and eventually thousands of miles would separate producers of cattle and other domestic animals from the locations where they were slaughtered and even further removed from the tables of distant consumers.[9]

Changes in the consumption of material objects and food transformed the many ways in which people related to each other and to the changing societies in which they lived. The consumption described here came into existence before the rise of capitalism and may have been responsible for its widespread acceptance as more discretionary income became available to more consumers. The eighteenth-century proliferation of early consumer goods in furnishings, mirrors, jewelry, cutlery, printed writing, pictorial prints, and newer fabrics will become apparent as this narrative unfolds. Staple and luxury foods arriving from distant places added meaningfully to this societal transformation.

This eighteenth-century explosion in consumption, initially among families with status but later among the middle and skilled working classes as prices declined and manufacturing flourished, was motivated by a social competition focused on material possessions. "What men and women had once hoped to inherit from their parents, they now expected to buy for themselves. What were once bought at the dictate of need, were now bought at the dictate of fashion. What were once bought for life might now be bought several times over."[10] What once took place in a small way now became an all-consuming activity as shopping for goods that formerly occurred at fairs, marketplaces, or by purchasing from peddlers now became available in networks of shops.

In eighteenth-century Britain, advertising became one of the more visible methods of selling the full range of new goods. To sell clothing and furnishings, hawkers (salesmen) fanned out from the cities to the provinces, enticing buyers to consume newer material goods. Fashion magazines, advertising columns in print, and trade cards distributed by these same hawkers had a similar intent. Obsolescence followed fashion. The "world of goods," their replacement, and a seemingly never-ending quest for more and better goods entered the fabric of society's cultural and economic framework. In the centuries to follow, consumers would invest more time and money in search of the ultimate purchases. In so doing, the impact of consuming merchandise – the travel, the packaging, the disposal of used material goods on the environment, initially ignored – would become as visible as the merchandise itself. By the nineteenth century changes that began to appear first in Britain had spread across modern Europe and infiltrated the market economy of the United States.

Material Goods

The consumption of tea, coffee, sugar, and tobacco grew rapidly in the century before 1750 CE, a date signaling the beginning of industrialization for many historians. As a result, an economic link between production and consumption became a feature of the larger narrative of a world consumed by the demand for goods. During the seventeenth century, the Dutch Republic had become the most active and richest maritime trading nation in the early modern world. Its relationship with the leaders of the Mughal Empire in India, mediated by the Dutch

East India Company, gave it access to goods and merchandise in high demand in European markets.[11]

As a state chartered company, capitalized by investors purchasing stock, the East India Company established permanent and fortified trading stations in Java, on the Coromandel Coast of India, and throughout South Asia. A brisk trade was carried out, including the exchange of Indian cotton textiles and Persian silks for Japanese silver in Nagasaki. Pepper and other spices entered the European market from Dutch sources. By the 1650s, the Company sent 20 ships annually to South Asia with 10,000 tons of supplies, money, trading goods, and people. An enormously valuable cargo returned each year.[12]

The impact of this wealth on Dutch cities was felt almost immediately. Brick began to replace wood and lime in home construction for reasons of status and to reduce the danger of fire. Interior designs that included more private space and larger interiors for display became more common in the homes of the wealthier burgher class. A century later, the homes of the Dutch middle class included distinct drawing, dining, and bedrooms. Long before industrialization, a compulsive acquisition of material things had taken hold among the wealthy and the upwardly mobile in Dutch society.

As the Dutch historian Jan de Vries has pointed out, the national wealth trickled downward as peasant farmers began to acquire mirrors, paintings, books, and clocks during the two centuries from 1550 to 1750 CE. In addition, they upgraded their furnishings by replacing simple square and rectangular tables with octagon-shaped and round tables; side-armed chairs replaced simple ones. Wooden storage boxes gave way to oak chests and wooden bowls and dishes were replaced with signature blue delft pottery. Window curtains became commonplace among all classes by 1650 CE, another sign of growing domesticity, privacy, and material improvements.[13] Of course, all of these manufactured luxury items required the use of increasing amounts of specialty woods, including mahogany, oak, cherry, and cedar. Dwindling forests reserves, increasing populations, greater fuel needs, and land clearing for agricultural expansion accentuated European consumer demands.

With each passing decade in the seventeenth century, larger numbers of farmers, artisans, tradesmen, craftsmen, merchants, and others had access to more goods. From 1675 to 1725 CE, those areas of England linked to London by transport or providing coal to the city received urban goods such as window curtains, clocks, books, china, pottery, mirrors, and pictures. These changes in the material culture of the hinterlands also changed behaviors of one kind or another. The availability of forks, knives, and earthenware replaced wooden bowls and tin utensils, suggesting gradual changes in eating and drinking habits. The organization of household space also points toward changes in behavior and changes in relationships.

Kitchens, as a separate unit in the household, became more common as materials for cooking, eating, and congregating became ubiquitous. Before the advent of the kitchen, cooking took place in a common living area. All the food was placed on the table at once; there were no separate courses. In many rural and urban households, all used a common vessel and not individual plates. Eating was accomplished with a knife and one's fingers. Knives and forks, uncommon at the beginning of the eighteenth century, became more available by 1750 CE. Likewise, wooden and pewter dishes and expensive tinware in kitchens gave way to decorated pottery and porcelain and much cheaper earthenware by mid-century. For the wealthy and middle classes, decorative silverware became a symbol of status rather than functional tableware.

One explanation for the availability of material goods for the masses was the drop in prices that occurred from the late 1700s to the late 1800s as capital investment during industrialization spread from Britain to the European continent and to the United States. In Britain, the price of broadcloth dropped as much as 50 percent during this 100-year period and another 20 percent in the decades that followed. The price of flannel dropped in a similar fashion while fustian, a linen and cotton blend, fell 75 percent. At these prices, consumers invested more of their income in textile products. They owned more clothes and a greater variety of apparel. As prices declined and fashions changed, consumers abandoned heavier and bulkier woolens for more popular, ready-made, less expensive and lighter woolens, linen, and cotton clothing.

In France from 1725 to 1785 CE, a revolution in consumption followed the pattern found earlier in Britain from 1665 to 1725 CE. Cheaper versions of luxury goods became available to the middle and lower classes of small shopkeepers, journeymen, day laborers, and domestic servants. Historian Cissie Fairchilds popularized the word "populuxe" to identify these commonly available items. As one European traveler wrote, he remembered seeing a stonecutter remove his fashionable coat, muff, and snuffbox before working on a piece of marble.[14] In fact, Paris not London became the fashion capital of Europe during this early modern era and has maintained a position of prominence in fashion into the twenty-first century.

As in Britain, bookcases, chests of drawers, writing tables, desks, tea tables, and gaming tables along with beds, bedside tables, and cane chairs furnished the most modest middle- and lower-class French apartments. Kitchens also changed in the ways noted above. With the production of lighter and cheaper textiles, cloaks, shirts, underwear, stockings, and umbrellas became available to the masses of urban dwellers. Unlike British consumers, however, the French had populuxe items, the inexpensive versions of luxuries such as fans, snuff boxes, and umbrellas, to allow them, if only momentarily, to indulge their fantasies about the benefits of upper-class status. When it came to watches, however, gold and silver ones won over the cheaper versions available to journeymen and servant classes. As an upper-class symbol of status, the gold watch knew no substitute among the French urban masses.

In Britain's North American colonies, buying material goods from the mother country rose steadily during the 1700s but picked up rapidly after 1740 CE. In fact, it grew more rapidly than the population that doubled every 25 years. Easy credit, low prices, and a growing consumer market for British textiles, furnishings, and earthenware made goods available to all classes of colonials. Even though colonial elites purchased the finest and most expensive items, few remained on the outside of the consumer boom. As reported in colonial Boston, Mohawk sachems "had all laced hats, and some of them laced match-coats and ruffled shirts [and] appeared 'a la mode François.' "[15] Fashionable clothing and a host of products ranging in grade, quality, color, and price captured the attention of increasingly sophisticated buyers. Becoming more attentive to details, consumers made demands for more variety and quality on merchants, who did the same to their suppliers. An economy in goods that would have been unheard of a generation before in the colonies now included "tea kettles, wearing linen, silver spoons, coats and hats, table cloths, iron boxes, a coffee mill, a pewter basin, a pair of stays, a calico gown, a necklace, a silk waistcoat, a scarlet long cloak, a camblet cloak, a blue cloth jacket, a pair of black silk stockings, and two pairs of pumps."[16]

As historian T. H. Breen has pointed out, consumption was about more than buying material objects. An unprecedented variety of material goods created a new visual imagery for colonial consumers who saw advertisements in newspapers and handbills and received on a daily basis visits from peddlers and itinerant merchants. A new language of styles and textures entered their vocabularies as they made choices and decisions about their person and possessions. They used this new language to express their aspirations and expectations.

Exciting possibilities appeared as they reinvented themselves according to the styles of the time. This was the message of a pamphlet about attire published in Philadelphia in 1772. Titled *The Miraculous Power of Clothes and the Dignity of Tailors, being an Essay on the Words, Clothes make Men*, the essayist noted that while visiting a tailor shop he "found [the craftsman] amidst a chaos of velvet, brocade, and other rich stuff, out of which he created illustrious personages, graces, honours, and other worthies," The message was clear that consumption created new realities and new expectations. The world around you was adaptable to new meaning and ideas about equality and the existing social order.[17]

Luxury Foods Become Commodities

All of the material goods possessed certain identifiable characteristics. They were objects to behold, admire, and use. They possessed a kind of permanency in two regards. They were located in a space with functional and decorative attributes. Even when they lost their appeal through wear or breakage and were discarded, they remained in our memories. Food possessed none of these qualities but remained more vital to our wellbeing than any of these material goods. Beginning around 1650 CE, edible luxuries became everyday consumable foods for increasing numbers of people, marking the first time in world history that such a transformation had taken place.

During the seventeenth and eighteenth centuries, sugar, tobacco, and three stimulant drinks, coffee, chocolate, and tea, entered the mass market. In the process millions of acres of virgin lands, including rainforests and woodlands, were put to the ax and transformed into plantations, farms, and fields for the production of mind-altering stimulants. Their popularity complemented and sustained the emerging factory system as household industries and small rural workshops faded away. To remain alert and productive in this newly emerging system, workers in textile and iron mills drank cheap stimulant drinks of coffee, tea, and cocoa. In a new industrial system characterized by its routine and monotony, beer soup and other concoctions caused drowsiness. As the need to remain alert and disciplined became the hallmarks of working-class life, they disappeared from workers' diets.

Tobacco

About 50 years after Columbus's first voyage, Portuguese royalty in Lisbon was introduced to tobacco. Sixty years later, it was growing in Belgium, Spain, Italy, Switzerland, and England in very small amounts. By 1600 CE, cultivation had spread to the Philippines, India, Java, Japan, West Africa, and China by European mariners. Chinese merchants carried tobacco to Mongolia, Tibet, and eastern Siberia. As a result, tobacco's consumption was nearly global a century after Columbus's first voyage. And unlike the other exotic substances such as sugar, coffee, chocolate, and tea, all social classes became addicted to tobacco quickly.[18]

The initial appeal of tobacco in Europe spread rapidly. In 1603 CE, the first year that we have accurate data, England imported 25,000 lb. of Spanish Orinoco tobacco. The Virginia Company created in 1606 CE, intending to extract natural resources and market them as commodities, established England's first colony of Jamestown in 1607 CE. In July 1613 CE, six years after the settlement of the Virginia colony and after many seasons of planting, experimenting, and harvesting tobacco, John Rolfe succeeded in developing a hybrid by combining tobacco from Trinidad with local Virginia leaf for the growing European market. This genetically engineered, highly addictive substance triggered a European boom.

Throughout the 1620s, high prices brought more Virginia Company tobacco growers into the market. Plantations spread along the tidewater region emptying into the Chesapeake Bay. A boom market in tobacco overwhelmed the Virginia economy and turned it into a colonial monoculture. Virginia became a "factory for turning topsoil into tobacco."[19] As was the case in any market-driven economy, the surge in supply quickly outdistanced European demand. Prices collapsed as quickly as they had risen previously. Tobacco at three shillings a pound dropped as low as a penny a pound. By 1628 CE, Virginia's export of tobacco to England reached 370,000 lb. At these reduced prices, this luxury drug became affordable to a new and much larger population of urban and rural users. Slow growth over the next two centuries allowed more and more ordinary consumers to enjoy the pleasure of consuming tobacco products and become addicted.[20]

Tobacco's primacy in the settlement of the New World led to its cultivation by the Spanish, English, French, and the Dutch in Venezuela, the Caribbean, and in the Chesapeake colonies. As heavy consumers of tobacco, settlers learned early on that it could be grown in humid and temperate climates and in a variety of soils. Unlike other crops, its short growing cycle meant that dried and cured tobacco was ready for the user's pipe within nine months after planting. Under these conditions, no tobacco crop possessed similar qualities in taste and texture but all contained the highly addictive nicotine compound. So, despite precipitous price declines in the decades after its introduction, profit margins for growers remained high.

As John Pory, a Jamestown planter, reported in 1619 CE, an individual working alone made a profit of 200 pounds annually, while another with six helpers made 1,000 pounds.[21] A century later 19 thousand tons entered the ports of England, mostly from its New World colonies.[22] By 1710 CE, Western Europe consumed about 70 million tons of tobacco and by the end of the century it rose to 120 million tons. By the eighteenth century, tobacco had become the first mass-consumed food-like commodity in Europe. It would remain one of the last to become mass produced as many small producers met the demands of consumers for pipe tobacco and snuff (a tobacco-based inhaled substance) for centuries.

To support tobacco use, clay pipe production spread with consumption. Initially, the enterprise was confined to London but it spread outward to the towns quickly and then to Germany, Scandinavia, and especially to Dutch pipe makers in Gouda. Manufacturing both simple and elaborately designed pipes accelerated with increased smoking. In the Netherlands about 15,000 pipe-makers worked in the industry with half of them concentrated in Gouda.[23] The environmental effects of concentrating the activities of artisans in a few urban settings resulted in more airborne pollutants as kilns emitted smoke and its related toxins in the manufacturing of hardened clay pipes. The extraction of clay from surface and subterranean deposits scarred the land in some locations permanently.

The transition from pipe smoking to cigars and cigarettes is obscured by regional and national folklore but Spanish chroniclers reported that Native Americans in Central and South America rolled tobacco into leaves of vegetable matter before ignition. By the seventeenth century, Spanish settlers had begun wrapping tobacco in paper, calling it a *papelate*. During the early decades of the nineteenth century the *papelate* was exported to France from Spain where it became the French *cigarette*. There, a state monopoly in tobacco sold 7,000 lb. in 1845 CE to local cigarette producers, an amount sufficient enough to make about three million cigarettes.[24]

Since the French preferred the non-acrid American tobacco to its own leaf, the adoption of American tobacco as the standard spread rapidly from France to Germany and to Russia. Once there, a mixture of American and Turkish tobacco entered the market. Production in Britain followed the Crimean War (1853–6 CE). Although the connection between the War and British tobacco use has never been documented, manufacturing cigarettes did not begin in Britain until after the War. Exports from Britain to the United States began after the American Civil War (1861–5 CE) and by 1869 production there reached two million cigarettes. Soon thereafter, Greek and Turkish immigrants to the United States blended Turkish with the more acidic American leaf to make cigarettes easier to inhale. Chewing tobacco, an American invention that combined tobacco with molasses, also rose rapidly during the post Civil War era. With a steady and uniform release of nicotine, the appeal of cigarettes grew rapidly as nineteenth- and twentieth-century consumers became addicted.[25]

The introduction of cigarette-rolling machines in the summer of 1885 CE caused production to jump from 9.8 million in 1881 to 744 million two years later.[26] With brand advertising, the industry consolidated into an oligarchy of three major manufacturers to achieve economies of scale. American Tobacco, Liggett and Myers, and P. Lorillard manufactured the three major brands of cigarettes, *Lucky Strike*, *Camel*, and *Chesterfield*, respectively, with a combined 70 percent market share that surpassed 268 billion cigarettes in 1950.[27] Targeting women in the 1920s as potential consumers helps to explain some of the increase. Addiction, however, created its own momentum and by 1987, the most popular selling brand, *Marlboro*, sold 293 billion cigarettes worldwide.

The history of tobacco use is an unfinished one, however. In addition to the hundreds of toxic chemicals found in cigarette smoke, including arsenic, cyanide, and nicotine, radioactive polonium 210 is one toxin absorbed by the tobacco plant from the uranium found in most soils. Using high-phosphate fertilizers to stimulate tobacco plant growth increases the transfer of uranium from soil to plant. For each smoked cigarette, a person inhales about 0.04 picocuries (a fraction of a trillionth of a curie which is a measure of radiation) of polonium 210. Inhaling a pack-and-a-half of cigarettes each day for a year is equivalent to 300 chest X-rays.[28]

The link of smoking to cancer, cardiovascular disease, macular degeneration, and a host of other illnesses resulted in significant declines in consumption among the middle class in the industrialized western world during the last decades of the twentieth century. The overall picture globally is not encouraging, however, with increasing numbers of smokers in the developing world and among the working classes in industrial countries. With about 5.7 trillion cigarettes consumed annually, mortality figures say all we need to know about tobacco's lethality. The World Health Organization estimates that 10 million people will die annually from smoking by 2020. Globally, 100 million deaths were attributed to smoking in the twentieth century. As smoking spreads rapidly through the developing world in the decades to follow, a pandemic in smoking-related deaths seems likely. The human costs in terms of the loss of life are staggering, with an estimated one billion people dying of smoking-related diseases in the present century. With it, medical costs and the loss of productivity will exceed hundreds of billions of dollars annually.[29]

The ecological impact of tobacco production and curing degrades the land, depletes woodlands, and spreads environmental devastation. Flue-cured tobacco accounts for most of the tobacco found in cigarettes. Until the past 40 years, American tobacco growers cured tobacco in closed buildings with wood-burning furnaces that directed heat from flues or pipes that extended from furnaces into barns. The temperature of furnaces was gradually raised from 90 °F to 160 °F until the leaves and stems were completely dried. With the heat and humidity controlled, moisture was removed, resulting in dried yellow leaves and stems. Most flue-cured tobacco used in the production of cigarettes has high sugar content and medium-to-high nicotine content.

The following data provide some perspective on the environmental costs of tobacco production and curing. Processing tobacco by this method required the cutting of one tree for every 300 hundred cigarettes. Estimates for curing a kilogram (2.2 lb.) of tobacco run as high as 12 cut trees worldwide today to a figure of as low as three trees cut. In developing regions of the world where this lower calculation was made, deforestation was identified as a persistent problem, however. Initially caused by the advance of modern agriculture, tobacco growing now plays an important role in rainforest decline.[30]

As woodlands around tobacco plantations deteriorated, American farmers growing flue-cured tobacco began to cure leaves through direct-fired systems that burned natural gas. This method has been found to produce tobacco-specific nitrosamines. Nitrosamines are produced by nitrous oxides, a product of combustion that combines with nicotine in tobacco leaves. Tobacco-specific nitrosamines have been shown to be carcinogenic, adding to the lethality of cured tobacco. Current research suggests that by retrofitting direct-fired systems with heat exchangers, tobacco-specific nitrosamines can be dramatically reduced but not eliminated. In an industry that sells a product that endangers users, those around them, and causes ecological devastation, retrofitting remains a distant plan.

Sugar

Merchandise of all kinds was exchanged across the global trading networks that linked Eurasia by way of the Silk Road, the Indian Ocean, and the Mediterranean Sea. Similarly, the Atlantic Ocean connected Europe and Africa to the Americas. As described by anthropologist Sidney Mintz, what had "lately [been] the luxuries of the leisured and rich, [became] the daily necessities of the overworked and poor."[31] The overworked and poor were the world's hungry and their need for staple food grew during these centuries, making the transformation of these luxury foods even more desirable to masses of people. Cheaper than beer that required the expensive distilling of grains, tea was called pejoratively that "oriental vegetable" from China. When sweetened with the Atlantic trade's West Indian sugar, it became Britain's most important beverage for day laborers, servants, and the poor.

High-priced sugar entered the English marketplace before the 1600s as a special treat usually offered as a gift. With a dramatic two-thirds cut in price in the sixteenth century and another 50 percent drop in the seventeenth followed by another one-third from 1700 to 1750 CE, sugar began to compete with meat and beer in the number of calories provided to the general population. By 1775 CE, per capita consumption of sugar reached 24 lb. per year. Some of the increasing demand for sugar as a sweetener can be explained by the growing popularity of caffeinated tea and other caffeine drinks. Although exact figures for Britain's North American colonies are not known, sugar consumption including that of molasses, a cheaper substitute, and used in the distillation of rum and rum punch, became a popular alcoholic beverage among the British at home and in the colonies. We know that their imports into Britain contributed 140 calories daily per person, while the colonists, conspicuous consumers of sweets, added 260 calories per person.[32]

As an all-important characteristic of human diets, the cultivation of sugarcane had a revolutionary impact on European nutrition and consumption. Agricultural societies thrived on the consumption of complex carbohydrates such as maize, potatoes, rice, millet, or wheat. Humans receive nourishment by converting these foods into body sugars. Other plant foods and fish, flesh, foul, fruit, berries, nuts, and honey are important but secondary to the core carbohydrates and their natural sugars. At the time when sugar became available to the British, they, along with most of the world's population, struggled to acquire and maintain a diet based on a single stable carbohydrate. As recently as the nineteenth century, most of the world's population lived on a single starch food (e.g., maize, wheat, potatoes, etc.) with other foods serving as a supplement. Hunger was a constant fear and outbreaks of famine were common.[33]

Before 1750 CE, the wealthy consumed most of a country's imported sugar. Limited production, taxes, and price made it a luxury food for the elites. With a century of increasing production, the elimination of import taxes, and falling prices, sugar became a commodity for all social classes and especially for the poor whose diets lacked protein and complex carbohydrates. By 1850 CE, sugar had become a daily commodity increasing the average total caloric value from about 2 percent of the total to an estimated 14 percent in the twentieth century.[34] In Britain, pastries, puddings, cheap jam-covered breads, biscuits, buns, other sweetened baked goods, and candy added calories to a nutrition-starved diet. Made with flour, baked goods complemented complex carbohydrates. Drinking sweetened tea, coffee, and chocolate along with baked goods became common during mealtime, for snacks, and at bedtime. The popularity of desserts, spread with the publication of a confectionery cookbook in 1760 CE, only added to a population's cravings for sweets.[35]

By 1800 CE, the European consumption of sugar reached 245,000 tons, most of the world's total production. Thirty years later, with sugar beet entering the market, production exceeded 570,000 tons, a 233 percent increase. With sugarcane and beet production accelerating by the end

Figure 8.1 Grinding sugar cane in a windmill. *Source*: Reproduced by permission of the British Library.

of the century, European and United States consumption reached more than six million tons of sugar. According to Mintz, "it may be enough to say that probably no other food in world history has had a comparable performance."[36]

As the incomes of workers rose slightly, the price of sugar declined as more producers entered the marketplace with a stable and growing supply of enslaved people from the West African/Atlantic slave trade. The sugarcane fields of the West Indies and Brazil flooded European markets with their product in the second half of the seventeenth century. First, it was used as a medicine, then as a spice, a preservative, a beverage sweetener, a pastry ingredient, and finally as a food. By the nineteenth century, sugar provided 15 to 18 percent of the calories for the entire nation. For some groups, probably the working and poorer classes, low-cost sugar may have represented a larger proportion of their daily calories. Some researchers argue that Britain experienced a twenty-fold increase in sugar consumption from 1663 to 1775 CE, while its population increased from 4.5 to only 7 million.[37]

Taste alone, however, cannot explain the widespread use of sugar. Along with the additional factor of a cold and rainy climate, hot tea laced with a sweetener warmed one's body. Other factors also played a role, including poor nutrition among urban workers. Mimicking the tea-drinking behavior of the upper classes may also explain its popularity. In addition, we need to think about the social and cultural effects of workers gathered together for the purpose of brewing tea in a shared teapot over a tinderbox to enliven a meal of bread and cheese. With each teacup they engaged in animated conversation that was enhanced by the taste of a hot and sweet beverage. These new products gave working men and women access to consumer products coming from a world separate from their own for the first time in world history. The production, processing, shipping, marketing, and consumption of the world's first modern commodities, tobacco, sugar, and the three drinks, tea, coffee, and cocoa, changed the relationship between producers and consumers and the relationship became irreversible.[38]

Despite the addictive attraction of tobacco, no product transformed the world of consumption like the production and distribution of sugar. Growers shifted production from other commodities, typically tobacco growing to sugar. Arriving on the European market in 1640 CE, it quickly became England's most consistently profitable American cash crop. From 1600 to 1800 CE sugar production became "the single most important of the internationally traded commodities, dwarfing in value the trade in grain, meat, fish, tobacco, cattle, spices, cloth and metals."[39] As the only consumer commodity responsible for transforming human relationships, agricultural production, and global dietary regimes, the "sugar revolution" was truly spectacular and at the same time intensely sad.

The sugar revolution transformed a diversified agriculture with mixed husbandry into a sugar monoculture that led to deforestation, soil erosion, and a landscape more vulnerable to the volatile weather systems of the Caribbean region. Large plantations ruled by a planter aristocracy, controlling an African slave labor population, replaced small farms owned by yeomen and supported by the labor of white indentured servants.[40] Sparse settlements of farmers gave way to dense, concentrated plantations. The plantation was "an absolutely unprecedented social, economic, and political institution, and by no means simply an innovation in the organization of agriculture."[41]

A diversified array of field crops and pasture along with habitat for insects, birds, and mammals disappeared as forests were clear-cut for cane fields. Planters cut nearby woodlands to boil away the raw cane juice and for cooking food. Roads and eventually railroads carved their way into the interior to transport crystallized sugar to refineries. From there, worldwide markets gained access to previously inaccessible lands transformed by capital investment in commodities and an evil labor system. An interconnected trade in sugar and its byproducts, molasses and rum, combined with timber from New England and Canada for the construction of Caribbean housing, and for shipbuilding for merchants engaged in the sugar trade. Looking for protein for their slaves, planters purchased tons of salt cod from Northern shippers who in turn bought it from international fishing fleets.[42]

The sugar revolution generated a booming Atlantic trade in slaves. Every European immigrant who arrived in the Americas from 1500 to 1820 CE was outnumbered by at least two African slaves. These slaves and their descendants worked the plantations, in one of the world's most ruthless systems of work, that created the New World wealth in the growing global mass-markets for sugar, tobacco, rice, and cotton. As early as the fifteenth century, African slaves provided the basic labor in Sicily and Naples. With the demand for sugar growing, European elites introduced sugar production and racial slavery in the Canary Islands and Sao Tome.[43]

Once brought to the Americas for the cultivation of sugar to satisfy this increasing European demand, African slaves suffered so severely that their birth rates were unable to maintain a stable population. Deaths exceeded births on sugar plantations in numbers not found on tobacco, cotton, or coffee plantations. In the New World, sugar was the great plantation crop, with from 60 to 70 percent of the slaves who survived the Atlantic crossing providing its basic labor. Life on these plantations was "nasty, brutish, and short."[44] In the Caribbean and Brazil where sugar crops flourished, slaves suffered population decreases of 20 percent each decade. Physical brutality, a tropical climate that served as an incubator of infectious diseases, and a poor ratio of males to females combined to make these plantations the source of great wealth for slave traders, owners, and merchants and dead ends for an enslaved workforce.[45] Only the repeated influx of new slaves from West Africa could sustain a sugar plantation economy. As Richard P. Tucker has pointed out, the unprecedented scale of production in the Caribbean represented, "the first flowering of American ecological imperialism."[46]

Coffee and Tea

"Theire ware also att this time a Turkish drink to be sould, almost in evry street, called Coffee, and a nother kind of drink called Tee, and also a drink called Chacolate, which was a very hearty drink."[47] Coffee drinking originated in the world of Islam where the coffee house was a place of sociability, economic exchange, and urban discourse. It grew in the valleys of Yemen between the Red Sea and the Indian Ocean. Along with spices, fabrics, and precious metals, coffee was a product of the thriving Indian Ocean and Silk Road trading network of the Islamic world. By the sixteenth century, its use had spread throughout the Ottoman Empire, including its largest cities – Constantinople, Alexandria, and Cairo – where Europeans lived and traded with Muslims. By the middle of the seventeenth century, coffee consumption had spread to London, Paris, and Amsterdam. According to historian John Wills, Jr., coffee houses in London became places where people went to read, write, debate religion and politics, and argue about constitutional principles. Coffee houses were the birthplaces of the modern urban polity.[48]

By 1700 CE, European demand for coffee accelerated trade with Mocha on the Yemen coast. Cultivation spread to the Caribbean island of Martinique in 1723 CE when a French naval officer, Gabriel Mathieu de Clieu, transported a single coffee plant from France to the island. Within five years, plants from Martinique were exported to Brazil, stimulating a boom in coffee cultivation and production in the western hemisphere. As demand began to outstrip the supply available from Yemen, Britain, with its colonial empire extending from the Caribbean into North America, opened coffee plantations in Jamaica and Barbados in 1728, while the Dutch, also in the 1720s, began production in Java. Through its controlling and centralized entity, the Dutch East India Company, production in Java surpassed Mocha by producing about five million pounds each year for the growing European market. Then, the company cut wages paid to its local producers and their peasant workers. Falling prices and a stimulant beverage were all that was needed for a growing European market.

By 1713 CE, the British had established regular trade with Guangzhou, China and stabilized its tea imports, making it a national consumer beverage. It took about 2 lb. of tea consumed each year for each person

Figure 8.2 Coffee shop in Palestine. *Source*: From Wikimedia Commons, the free media repository.

to have a cup of tea every day. Assuming that 25 percent of the adult population of 951,000 persons drank a cup each day, then the British needed to import 1,902,000 lb. each year. Every student in the United States knows the history of the Boston Tea Party (1773 CE) in which colonists dressed as Native Americans dumped East India tea into Boston Harbor because the British government imposed a duty on tea, a beverage as popular in the colonies as in Britain. Taxes on tea imports almost doubled its price, leading to a massive and lucrative international smuggling trade. It invaded coffee houses along with chocolate, giving customers additional stimulant beverages to drink. As prices for black tea called Bohea and green tea continued to decline, more and more drinkers entered the market. Soon, sugar bowls and milk pitchers appeared along with tea, creating a brew that was warm, sweet, and nutritious. Its popularity spread rapidly not only in Britain but in France and the Netherlands as well.

Figure 8.3 A tea warehouse in Guangzhou, China. Porters carry baskets of tea to three European inspectors. *Source*: Reproduced by permission of the British Library.

As stimulants, hot caffeine drinks of coffee, tea, and cocoa competed with cold beer and spirits for the attention of the masses. Contemporaries saw the impact of these mood-altering drugs on the alertness of workers and their attention to tasks. Laborers who formerly drank beer for breakfast and arrived at work sleepy now drank strong coffee sweetened with sugar and arrived awake, alert, and attentive. Although coffee and tea competed for a century or more in Britain for the loyalty of workers, in the long run tea won out. Unlike coffee, tea required no household grinding equipment; it remained the cheaper of the two beverages with one pound of tea making up to five gallons of fresh liquid. On the other hand, a strong cup of coffee required 2 to 3 oz of beans.

By 1750 CE, however, growers had located coffee plantations in the tropics and temperate zones of five continents. By the end of the nineteenth century, coffee drinkers in the United States would consume one-half of the world's supply. Nearly three-quarters came from Brazil, with coffee from Guatemala, Costa Rica, and Mocha from Arabia and

Java from Indonesia competing for the remaining quarter.[49] By 1900, the United States imported almost 400 million pounds. In the decades to follow, coffee consumption continued to rise, with the United States share mostly responsible for this increase.

At the beginning of the twentieth century, 167 million Americans drank coffee, with as many as 5 million new coffee drinkers entering the market each year. Later in the century, the rise of brand advertising, aggressive marketing, and the arrival and the spread of specialty chains such as *Starbucks, Second Cup, Dunkin' Donuts,* and the *Coffee Beanery* accounted for some of coffee's growing appeal. With more consumers entering the marketplace, coffee cultivation in Latin America reached record levels.

Oversupply resulted in driving the price of coffee beans to new lows. During the height of the coffee boom years in 1997, the retail prices of coffee reached $4.67 a pound. Despite increasing demand, six years of high yields caused prices to plummet to $2.84 in 2003. These new low retail prices increased demand even further, thereby continuing the push and pull mechanism behind coffee consumption in the twenty-first century. In the United States, the average consumer drinks more than four cups of coffee each day. Although this number represents more than half of the country's 300 million people, other nations consume more on a per capita basis. For example, almost all coffee users in Scandinavia, including Sweden, Finland, Norway, Demark, and Iceland, drink on average seven or more cups each day, with Finland leading the group with 11.5 cups per day.[50]

Environmental Effects of Increased Cultivation of Coffee

Almost two centuries ago, Charles Darwin observed: "The land is one great wild, untidy luxuriant hothouse, made by nature herself."[51] Coffee cultivation, along with intensive forms of agriculture and grazing animals, has long ago transformed that "untidy luxuriant hothouse" into ecologically fragile landscapes of managed fields for domestic plants of declining nutritional value. Like other crops, the use of pesticides to promote growth became widespread on coffee plantations. As the third most heavily sprayed crop on the planet, with only cotton and tobacco ahead of it, coffee cultivation requires the use of millions of gallons of petroleum-based chemical fertilizers to thrive. This is another use of a fossil fuel that consumers of coffee neither recognize nor acknowledge.

An important part of the process that ultimately brings coffee to the consumer requires the removal of the outer shell of the coffee bean in preparation for roasting. For generations of coffee growers, soaking the shell that protected the bean was the first step in the production process. This wet process produced tons of fermented waste that was flushed into rivers and streams causing widespread pollution. Rotting fermented waste depleted water of its oxygen and killed plant and fish life. Although this kind of degradation is widespread, newer methods of depulping without water have curtailed the spread of pollution and produced some positive environmental effects.

Instead of flushing the red-skinned pulp into adjacent waterways, it is placed in piles and mixed with lime, producing high-quality organic fertilizer for use on the fields of these same coffee plantations. This is only one kind of recycling. On other coffee plantations, the California red worm is introduced to the piles of pulp and within a few months transforms it into organically rich soil. Other forms of recycling include mixing coffee mucilage with cow and pig manure in an underground tank where it undergoes decomposition, producing enough methane gas to use for cooking and heating.[52]

Despite the ongoing hazards posed by the processing of coffee beans and the recent efforts to soften their negative environmental impacts, the ongoing assault on shade-grown coffee threatens a vital home for migratory birds. "Thousands of birds fill the air with song-pert green parakeets, big gray mockingbirds, brilliant bluebirds and little yellow canaries. It is difficult to imagine anything more delightful than a ride through the long avenues of trees heavy with green coffee berries. When new ground is to be planted in coffee, shade is the most important consideration."[53] Since the 1920s when this observation was made, the assault on shade-grown coffee has resulted in more than 70 percent of Central and South American coffee being grown in microenvironments of densely packed patterns in full sun. As coffee groves replaced traditional local forms of vegetation, the additional transition to a sun-filled coffee revolution from shade-grown coffee upset an ecological balance that included millions of migratory birds that nested in shade trees and consumed billions of insects in the form of coffee borers and other destructive pests.

The loss of important habitat near tropical landscapes to a wide variety of migratory birds, including swallows, warblers, orioles, thrushes, and hummingbirds, contributed to ecological degradation. During the last decades of the twentieth century, as many as 10 billion birds inhabited the summer climates of North America's temperate forests and flew to Central and South America during the winter months. With the increasing destruction of millions of acres of forest for cultivation, migratory populations of birds have declined.

In the last century, the world's natural wooded landscapes underwent massive changes. Aside from the forests in temperate zones, tropical rainforests are disappearing at an alarming rate. Once, they covered five billion acres or 14 percent of the Earth's land surface. Today, the wooded tropical land has been transformed by human action. Half has already disappeared and the rate of loss at approximately 80 acres a minute continues unabated. The extinction rate of species lost to land clearing for all purposes, as best we can estimate, appears to be about three species per hour. The once "wild and untidy luxuriant hothouse" of Charles Darwin is gone forever.

Conspicuous Consumption

The consumer revolution became the "magnetic center of society. Profound changes in consumption created profound changes in society and these in turn had created further changes in consumption."[54] The establishment of department stores focused the attention of consumers, created a pricing system for goods that eliminated bargaining between buyer and merchant, and invented new strategies for marketing material things. Designed to add status and value to goods, these new strategies included testimonials by celebrities, appeals to the aesthetic and cultural qualities of objects, and the use of models who represented a stylized and romanticized image for all potential buyers to emulate. The style of mass consumption introduced by the emergence of the department store was loaded with cultural meanings about status, taste, art, gender identity, and many others.

By century's end, the economist and social critic Thorstein Veblen wrote about "conspicuous consumption" in his *Theory of the Leisure Class* (1899) and argued that material goods became a way for all social groups to display their sense of success and to emulate society's elites.

Throughout much of the twentieth century, the opinions of writers varied from extolling the virtue of consuming and the leisure and recreation that it fostered to those who pointed out the obsessive and compulsive behavior of shoppers and the inequities caused by individual consumption at the expense of social consumption. Social consumption included schools, hospitals, roads, and other public venues that were neglected while individuals indulged in instant gratification.[55]

From Thomas Edison's first light bulb in 1879 to the opening of America's first power plant in 1882 that illuminated New York City's financial district with 1,300 light bulbs, the impact of electricity on the manufacture of consumer goods in the twentieth century would have been unimaginable. Two other inventions became vital additions in advancing the consumer revolution. High-voltage transformers transmitted electricity to low-voltage households and the electric motor became vital to industrial expansion in railroads, telecommunications, and households. By 1900, electric appliances were becoming household items – electric irons, vacuums, water heaters, and cookers entered many middle-class American homes. Their spread was limited only by the availability of household wiring.

The purchase of a home became the most significant consumer purchase in the post World War II period. The Great Depression of the 1930s, the war that followed, and a rising birth rate created a housing crisis in America. After World War II, government publications and advertisements from real estate developers promoted the ideal home as a single-family detached home filled with appliances. A building boom that began during the war years, with an unprecedented 15 percent increase in homes owned by occupants, continued almost unabated throughout the rest of the twentieth century. The home portrayed in the plans of builders was suburban in style. Similarly, magazines such as *Better Homes and Gardens* provided the eager consumer with house plans and decorating ideas.[56]

Expenditures for residential construction and consumer goods and services, rising to a total of 70 percent, drove the postwar economic recovery in the United States. One of every four owner-occupied single family homes in 1960 was built in the 1950s. In 1940 only 44 percent of Americans owned a home but 20 years later the figure had risen to 62 percent and it would continue to rise throughout the century. By the 1970s, 85 percent owned their homes, if the profile of the owner consisted of a two-parent family between the ages of 45 and 64 years old. As major metropolitan area growth stagnated, suburban growth almost doubled, with a steady growth of new homes built each year. By 1950, annual housing starts had reached two million units and stabilized at one million a year in the decades to follow.[57]

Household construction had an enormous ripple effect with the growing need for schools, places of worship, roads, and recreational facilities, including restaurants and resorts. With the electrical infrastructure of homes changing with the onset of the housing boom, refrigerators appeared in the 1940s followed by washing machines, air-conditioners, self-cleaning ovens, food processors, and microwave ovens. The spread of these disposable consumer goods among a growing middle class sparked spending and buying on credit for the first time. Appliances with new features, designs, colors, sizes, and pricing advertised by a print and a growing electronic media created market penetration across income and social class. How new one's possessions were and how much one owned distinguished purchasers by class.

Global Consumption

Further distinctions based on spending separated participants in the consumer revolution. Many houses and dwellings around the world either have no electrical service or only a few sockets, while modern households contain complex wiring and as many as 50 electrical sockets and switches.[58] The gap widened, as access to services required electronic communication and improved education. Many areas of the world remain pre-modern when measured by the standards of a modern consuming society.

In 1954, fewer than 60 percent of French households possessed running water. Only 25 percent had installed an indoor toilet and only 10 percent had central heating and a bathroom. By 1975, however, three-quarters of the households included an indoor toilet, 70 percent had bathrooms, and central heating was found in about 60 percent of the homes. By 1990, these appliances had become commonplace in French homes. Rising levels of consumption improved hygiene and expectations about the relationship between cleanliness and social standing in the larger community. A similar pattern can be found in other mass-consuming societies, with the sequence of events described here for France taking place somewhat earlier in some countries and later in others.[59]

Figure 8.4 The Mangga Dua Mall in Jakarta, Indonesia. *Source*: From Wikimedia Commons, the free media repository.

The revolution in transportation and the introduction of the automobile, itself a means to consume more resources and goods, meant that the automobile became a major consumer durable in countries with high economic growth rates and more disposable income. For thousands of years, transportation on foot or on the backs of animals moved people and products. Railroads began running in Britain in 1847 and transformed the meaning of time and space. Distances from one place to the next were shortened and soon after the invention of the railroad travelers rode in comfort. Meals, sleeping accommodations, and several amenities attracted more and more railway passengers. Freight rail cars carried natural resources, foodstuffs, and finished manufactured goods to distant markets, further separating consumers from the natural world of animal protein and plants.

In many countries, rail transportation continued to expand into the twentieth century, but by the 1950s motorized vehicles and planes began to compete with rail traffic for passengers and freight in industrialized countries. Economic, social, and environmental changes brought about by the automobile rank as among the most far-reaching transformations in the modern era.[60] By the 1920s in the United States and in other industrialized countries soon thereafter, manufacturing automobiles became big business. They became major commodities for both domestic consumption and international trade in the United States, Germany in the 1960s, and Japan in the 1970s. Quadrupling car sales in the decade from 1946 to 1955 resulted in three-quarters of households owning at least one automobile by the end of the decade. By 1975, 85 percent owned a car and almost one-half of all households owned two cars. Peak production was reached in 1965 when the industry built 11.1 million cars, trucks, and buses in the United States.[61]

With the movement to the suburbs and the relative absence of interurban railroads, the automobile became both a necessity and a highly valued consumer choice. Fascinated with cars, their mechanical power, and highly varied size and styles, consumers viewed them as symbols of status and recognition. They also proved to be a boon to the oil industry, service stations, dealerships, repair shops, parts suppliers, and the growing travel business. Motels, restaurants, resorts, and other recreational facilities grew up along the many thousands of miles of roadways built to accommodate the increasing numbers of cars. In 1937, seven shopping malls dotted the landscape; in 1949 there were 70 and five years later, one thousand. With accessible roadways, highways, and free parking, their numbers continued to grow. By 1960, 4,000 shopping centers and malls were providing suburban households with department stores, specialty shops, restaurants, and recreation. In the twenty-first century, thousands more malls are operating, with new ones appearing almost daily.[62]

As the car became a high economic value consumer purchase along with the single family home, it stimulated growth in other industries. Steel, rubber, glass, plastics, upholsteries, and electronics grew, with auto manufacturing linking one out of every six jobs to this growing consumer market. They declined when the United States auto industry began facing competition for its vehicles at home, first from Germany in the 1960s and soon thereafter from Japan whose manufacturing plants now produce more automobiles than any other country. The world of automobile manufacturing had changed significantly since 1927 when 85 percent of the world's cars were manufactured in the

United States. Much of the country's industrial capacity was tied so closely to the phenomenon of the car that America's decline in the manufacture of automobiles reverberated throughout the economy.

The automobile represents the extension of private space from the household to a place of employment, recreation, and pleasure motoring. It liberates most people, giving them greater access to jobs, recreation, and the world around them. Excursions into the countryside from cities redefine urban space by linking new suburban communities to larger metropolitan areas. To accommodate this truly intrusive vehicle, roadways of all kinds, highways, freeways, and toll-ways, cut across large residential neighborhoods and commercial zones dividing each from the other. The city, as an integrated mosaic of multi-functional neighborhoods with mixed housing, small retail specialty shops, and larger department stores, was fractured into isolated enclaves and in time in some places became depopulated and abandoned space.

The Automobile and Electronics in Emerging Markets

Unlike mature and saturated markets for cars in North America, Europe, and Japan, "new automobile spaces," or emerging markets for automobiles, exist in Southeast Asia, Central Europe, China, India, Argentina, Brazil, and Mexico. Since the 1990s, these regions and countries have become areas of growth for car sales and ownership. Along with that growth in consumption came vehicular-driven atmospheric pollution in many countries without emission controls in the form of catalytic converters, electronic ignition, computer controls, unleaded gasoline, and a high percentage of aging vehicles. The exposure to air pollution by children in some South Asian cities was equivalent to smoking two packs of cigarettes a day.[63]

In China, for example, low emission standards mean that its vehicles pollute more than any others in the world. Although leaded gasoline is currently being phased out, Beijing creates more air pollution than the combined total for Los Angeles and Tokyo which are 10 times larger than the Chinese capital. Without the modernization of traffic infrastructure to keep vehicles moving, the use of newer technologies, and auto emission standards, emerging market countries like China will surpass the total greenhouse gas emissions of developed countries in two to three decades.

Currently, the combined energy consumption in China and India represents half of that consumed in the United States. As their personal incomes and levels of consumption rise and they begin to manufacture and import more vehicles, their emissions will rise as well. In 2000, United States per-person ownership of motor vehicles exceeded that of China by a factor of 127.[64] As a result, they and many other countries have a long way to go before they approach consumption levels in the United States.

CHAPTER NINE

FOSSIL FUELS, WIND, WATER, NUCLEAR AND SOLAR ENERGY

Introduction

In *Technics and Civilization* (1934), Lewis Mumford divided the use of energy by humans into three distinct time periods: the eotechnic (the prefix meaning "dawn"), the paleotechnic (the prefix meaning "old"), and the neotechnic (the prefix meaning "new"). Although the eotechnic represented the longest historical period in which human and animal muscle power provided the kinetic energy to perform work, Mumford's description focused on energy that came from capturing the wind, burning biomass (primarily wood), and harnessing running water. The different periods overlap in many ways, for reasons that will become clear while reading this chapter. Despite the overlap, however, Mumford's classifications remain a useful paradigm for studying the impact of energy use on the global ecosystem.

Turning water into steam to power engines and mining and burning fossil coal to power machinery signified the entry of humans into the paleotechnic world. The substitution of oil and natural gas for coal to generate electricity for machines, vehicles, households, and appliances of every size and shape reflected an expansion of the paleotechnic into everyday life and a transition to the neotechnic age. The neotechnic substituted oil and natural gas for coal in generating much of our electrical needs, while the internal combustion engine replaced the animal power of the horse.

As with almost every paradigm, however, the exact timing of the transition from one period to another remains a subject for debate. Many activities and their mechanical aids are carried over from one period to the next. The ax (an eotechnic implement) remained the main instrument for clearing the land of trees in preparation for farming at the same time that railroads (a paleotechnic invention) carried people, produce, and natural resources to and from frontier regions.

Within a mature neotechnic world, the burning of fossil coal remains the most plentiful resource in generating millions of kilowatt-hours (kWh) of electricity. The United States alone burns more than one and a half billion tons of coal annually to generate electricity. The global consumption of coal rises annually, producing a large share of the world's carbon emissions and their contribution to global warming. At the same time, nuclear power and hydropower combined contribute more than 20 percent of the nation's electricity production.

The Eotechnic World: Waterwheels and Windmills

People faced numerous hardships during the eotechnic. The extremes and vagaries of the climate placed severe limitations on human activity. During extended periods of intense cold, a shortened growing season made daily hunger a constant threat to public health. Rivers and streams froze, making passage by boat impossible. Frozen harbors imperiled commercial exchange and added to the economic uncertainty posed by an unpredictable climate. Extended cold periods penetrated the household in ways unknown to later generations living in the paleotechnic and neotechnic worlds. Inefficient fireplaces sent most of the heat up chimneys and drew in cold from the out-of-doors. More efficient closed stoves burned less cordwood but their higher prices limited their distribution. As local supplies of wood dwindled and costs rose, poorer households faced increasingly lower indoors temperatures. They were not alone, however, because many rich households experienced similar indoor conditions. John Rowe, a late eighteenth-century Boston merchant, noted, "the ink freezes as I write."[1]

Cold temperatures during the eotechnic were not without their advantages, however. The tropical infectious diseases such as malaria and yellow fever seldom invaded frigid regions. The poor condition of roads everywhere in the world inhibited travel during periods of flood and drought. In the former circumstance, inundated roads prohibited travel, while in the latter, dusty, deeply rutted roads made travel a bone-jarring experience for persons and cargo. Snow- and ice-covered roads created low-friction surfaces suitable for the rapid movement of heavy loads of goods (including cordwood), reducing the amount of energy expended by draft animals and humans. Growing cities benefited greatly from supplies delivered regularly from the outlying towns and farms on winter sleds. "In the winter-time, when the ground was hard with frost or covered with snow, clumsy carts and sleds, drawn mainly by oxen, were kept busy bringing loads of cordwood from the woodlots, or carrying corn, potatoes and other farm produce to market at Boston."[2]

To gain some perspective on the length of the eotechnic period in human history, many countries continue to depend on muscle and water for power. High-energy societies based on the consumption of fossil coal, oil, and natural gas are only about 120 years old, meaning that most emerging modern economies depended on muscle and water for power during their initial development. During the American Revolution, the colonies possessed only three steam engines and these were used for pumping water. As late as 1840, wood provided more heat than coal and more than 60,000 of the country's manufacturing mills were powered by waterwheels. They provided much more energy than the 1,200 stationary steam engines that averaged 20 horsepower. Steam passed waterpower in manufacturing in 1875 but farm labor continued on an eotechnic energy path dominated by human and animal power. At that time, only one household in ten possessed electricity or a telephone.[3] These represent only a few examples of eotechnic, paleotechnic, and neotechnic functions coexisting simultaneously during transitional phases in human history.

In the meantime, between 1820 and the end of the century, horses, oxen, and other draft animals remained the predominant mode of urban transportation. Humans pulling small open coaches also operated in many Asian cities. Horse-drawn wagons, carts, and vans met most railway shipments upon arrival in the world's growing cities and distributed manufactured goods and agricultural produce to retailers and grocers throughout the urban landscape. As many urbanites accumulated wealth, they purchased private horse-drawn coaches and hansoms and became primary users of the city's expanding number of cabs. The urban working class traveled using the growing number of horse-propelled omnibuses, first introduced in London in 1829.[4]

At the same time, feeding horses consumed agriculture's largest supply of hay and oats, while stabling them required committing large urban space for this purpose. By the end of the nineteenth century, Londoners owned some 300,000 horses and New Yorkers, facing the burden of growing piles of manure on its streets, considered removing horses to a designated belt of suburban pasture during periods of low demand for transportation. Calculating the urban-energy footprint of horse-drawn transportation in terms of feeding, stabling, grooming, shoeing, harnessing, driving, and waste removal remained one of the largest items in the energy costs of nineteenth-century cities. As the number of horses reached its peak at the close of the nineteenth century, electric trolleys and internal combustion cars and trucks began to replace them. In the developing world today, eotechnic forms of transportation impose a burden on growing urban populations similar to that faced by developed countries a century ago.

The dominance of an energy system based on animal muscle lasted for as long as it did in the West for many reasons. Breeding larger and more powerful draft animals such as the Scottish Clydesdale and the English Shire horses and technical improvements stand out as two of the most important innovations. Geographer Vaclav Smil has written about the ways in which animal power became more efficient and productive. "A heavy nineteenth-century draft horse hitched to a light, flat-topped wagon on a hard-topped road, equipped with iron horseshoes and a collar harness, could easily pull a load twenty times heavier than the load that could be pulled by its much lighter, unshod, breast-harnessed ancestor linked to a heavy wooden cart on a muddy road."[5]

The invention of mechanical tools in the form of wedges, screws, thread and gear wheels extended the world of the eotechnic. Machines and tools enlarged human control over objects larger than themselves. Lifting, grinding, crushing, and pounding materials in the natural world provided humans with a significant advantage in shaping that world to their own ends. Despite these advantages, however, the kinetic energy of animals and humans was still required to complete a number of labor-intensive tasks. Great wheels, powered by human kinetic energy turning a crank, allowed others to make precision wooden and metal parts on a lathe. Manufacturing steel parts for the first locomotives in 1813 required two strong men. According to the manufacturer, George Stephenson, the job was so exhausting that five minutes of work demanded an equal number of minutes at rest.[6]

Treadwheels, an innovation that lightened the load of workers, became an efficient device for raising water from a well, while its companion chain-and-ball pump lifted heavy building materials at construction sites. The building of great structures during the Middle Ages required innovative technologies to lift and precisely place large cut stones. Experimenting with the size and efficiency of treadwheels, gearwheels, pulleys, cranks, cranes, and pumps over decades and sometimes centuries of use to accomplish increasingly complicated tasks taxed human ingenuity. Solving problems not encountered before became the vehicle for experimentation, innovation, and human progress. During the transition to the paleotechnic era, many of these machines would become essential tools in subterranean mines for raising coal, ore, and water. Mine owners and operators employed the power of steam engines to drain flooded mines before that power was enlisted to move locomotives.

Long hauls, formerly the domain of oxen hitched to wagons weighted with loads impossible for humans to move alone, quickly disappeared with the invention of the steam engine and its application to the locomotive. Only the amount of energy produced by burning fossil coal, converting water to steam in the locomotive's boiler, and transferring the steam to its wheels limited the number of railcars coupled to a steam-driven locomotive. The load it pulled and its speed remained a function of this energy exchange. Animal and human power would not withstand the emergence of rail technology.

As with most paradigm shifts, however, the transition did not eliminate all eotechnic functions. Within the boundaries of most cities during the winter months, runners replaced wheels on local horse-drawn private and public carriages and delivery wagons. The energy provided by horses remained a fixture on city streets and a sanitation nightmare into the twentieth century in the developed world and remains one for modernizing economies. In the developed world, the transition was complete with the arrival of the gasoline-powered internal combustion engine that provided the energy supply for automobiles and trucks. Within a few short decades, the horse as a beast of burden disappeared from the streets of most Western cities, while around the world it remains an example of an earlier energy regime.

Innovation took many forms and the design of advanced waterwheels and windmills was a long time in coming. For millennia, the technology of both remained largely unchanged. The first reference to a watermill for grinding grain and the first windmills to raise groundwater to irrigate desert gardens appeared in Persia in 4100 BP. Despite the existence of wind-powered sailing ships, it took many millennia before the inventive uses of windmills became widespread land-based

Figure 9.1   The Maid of the Mill. *Source*: From Wikimedia Commons, the free media repository. Johnson, Helen Kendrik (Ed): "World's Best Music" (1900) [1] (http://www.fromoldbooks.org/WorldsBestMusic/pages/074-The-Maid0fTheMill/).

mechanisms for generating power. In the well-watered and post-glacial world of Europe, however, waterwheels dominated the landscape. Propelled by the kinetic energy of moving water, waterwheels were visible throughout many parts of Europe and, according to Vaclav Smil, "everywhere east of Syria." England's first great census, the Domesday Book (1086 CE) commissioned by William the Conqueror, recorded the existence of 5,624 mills in southern and eastern England. One mill provided mechanical power for every 350 people.[7]

These early wheels converted about 20 percent of the water's energy into usable power and by 1800 the conversion ratio had reached 35 to 40 percent. Further technological advances in the nineteenth century converted as much as 60 percent of the energy into power. These advances did not come without costs, however, since maintaining a steady stream and river flow could not be left to the vagaries of the weather. Storage ponds, reservoirs, races, and dams became the mechanisms for regularizing the capacity of the mills to do their work, despite the fact that much of this work until the Middle Ages entailed grain milling. Such alterations in natural stream flows disrupted the annual migration of spawning fish, destroyed fishing rights, caused silting that inhibited water traffic, and caused upstream flooding of pastures and farmland. Complaints from farmers about changing stream flow reverberated through the annual agricultural reports of farmers everywhere.

As early as 1640 in Braintree, Massachusetts, fishermen disrupted an ironworks by destroying its dam because they claimed that it prevented salmon, eels, and other species from migrating to freshwater spawning areas. Despite legislation to build fish ladders and fishways near dam sites, by the nineteenth century Atlantic salmon had disappeared from rivers and streams. In 1857, George Perkins Marsh, diplomat, ecologist, and Vermont farmer, reported that, "Almost all the processes of agriculture, and of the mechanical and chemical industry are fatally destructive to aquatic animals within reach of their influence [and] to fish which live or spawn in fresh water. The sawdust from the lumber mills clogs their gills, and the thousand deleterious mineral substances, discharged into the rivers from metallurgical, chemical, and manufacturing establishments, poison them by shoals."[8]

The naturalist Henry David Thoreau noted the simplicity of a mill's design as one explanation for the destruction of forests and for the more than 66,000 watermills operating in the United States in the 1840s. "How simple the machinery of the mill! Miles has dammed a stream, raised a pond or head of water, and placed an old horizontal mill-wheel in position to receive a jet of water on its buckets, transferred the motion to a horizontal shaft and saw by a few cog wheels and simple gearing, and, throwing a roof of slabs over all, at the outlet of the pond you have a mill."[9] Extended work schedules for millers, sometimes reaching a 16-hour day, became a function of the amount of water held in reservoirs. Again, Thoreau, hearing the sawmills working at night during the high water levels in May, commented: "I can imagine the sawyer, with his lantern and bar in hand, standing by, amid the shadows cast by his light. There is a sonorous vibration and ring to it, as if from the nerves of the tortured log. Tearing its entrails."[10]

With the manipulation of water flow, the expanded energy output of waterwheels released humans and animals from numerous arduous tasks. The following list illustrates the immediate reach of water-powered machines for: sawing wood, turning it on a lathe, making paper, manufacturing textiles, tanning leather, crushing ores, stamping, cutting, and grinding iron and other metals, making horse shoes, and many others too numerous to catalogue here. The expansion of mining and metallurgy using waterwheels spearheaded a revolution in manufacturing even though it would take decades and even centuries to alter the design features of most functional waterwheels to produce power greater than large draft animals.

When changes in machine design were implemented, their impact became apparent. In 1795, a United States patent was granted to a manufacturer whose water-powered machine cut and headed 200,000 nails a day. Its economic impact was staggering as the price of nails, a considerable cost item in the construction budget of any builder, was cut by 90 percent during the next 50 years. In Britain, the country's largest waterworks near Glasgow, Scotland, on the Clyde River, maintained 30 mills in 1845 that were fed by a large reservoir and produced about 1.5 megawatts (MW) of power![11] David Nye, historian of technology, has noted that technological intervention into the natural world destroys or drives away the biological world. The mill narrative of economic progress and building human communities around rivers and streams overlooks the destruction or displacement of fish in the watershed.[12]

As the pace of technological and economic change quickened, water turbines replaced large waterwheels. These new turbine designs became the workhorse of textile manufacturing in Massachusetts by 1875. Their combined power along the textile-producing corridor of the Merrimack River in Lowell and Lawrence totaled 7.2 MW and accounted for about four-fifths of all the power available to the mills. Although the history of waterwheels extended over the millennia, water turbines fell victim to an accelerated pace of industry and the insatiable demand for more efficient sources of power. By 1880 in the trans-Atlantic world, the paleotechnic world of mining fossil coal and using it to convert water to steam, in increasingly more efficient engines, displaced water power's position of leadership. But before we plunge headlong into the paleotechnic era, we must address the matter of the wind.

What is the wind? It represents a form of solar energy that envelopes the Earth as it makes one complete rotation on its axis each day. As the Earth spins, massive streams of air circulate, some reaching velocities as high as 300 miles per hour (mph). Since the Equator is closer to its energy source, the Sun, surface temperatures vary based on its distance from the Sun. Rising and falling surface temperatures cause moving currents of air that we call the wind. The great differences in temperatures account for the greater velocity of winds on a global scale.

The evolving topography of the Earth's surface shaped the global climate system. If the surface were smooth then the movement of polar air from the North and the South in search of warm equatorial air would become a simple calculation of cooling and rising tropical air. The Sun's energy heats the air differently, however, as it travels over bodies of water and land. The wind is a manifestation of this energy exchange. A global surface punctuated with deserts, grasslands, forests, mountain ranges, oceans and other large bodies of water add complexity to atmospheric circulations.

Since the Earth is a sphere with points on its surface making smaller revolutions as these points come closer to the Poles, wind speeds change with changes in the shape of the sphere. A point on the Equator travels eastward 25,000 miles in a single day at a speed of about 1,000 mph. Further north, a point travels eastward 20,000 miles at 800 mph. Wind currents traveling north from the Equator will be traveling faster than the surface velocities, as will winds traveling south. Winds traveling from the poles toward the Equator will be traveling slower than the surface velocities. A very complex system of highly variable and volatile winds is the result.[13]

The ancient mariners and shipbuilders were the first to understand this complexity. From the China Sea and the Indian Ocean to the Mediterranean Sea and the Atlantic Ocean, they built and sailed ships knowing how to use the clockwise systems of wind circulation on the oceans. The Chinese sent large trading fleets across the Indian Ocean to Africa's east coast before exploration became a passion for European mariners. The Portuguese and the Spanish set sail during the Age of Discovery in the fifteenth century across the Atlantic following the westerly winds of the temperate zones and returned eastward from the tropics on the prevailing easterly winds.[14]

Soon after the discovery of the Americas by Europeans, 100-ton sailing ships, with crews of 80 men, generated 500 to 750 horsepower and traveled at speeds of 10 knots per hour. Although it was possible to row across the Atlantic, the sailing ship connected Europe and Africa to the western hemisphere, bringing wealth in the form of gold, silver, lumber, cotton, and a variety of food stuffs to long-suffering Europeans. Africans received none of these benefits. As slaves from West Africa, they provided the human muscle power to enrich the wealthy planter classes, slave traders, and textile manufacturers on both sides of the Atlantic. Boston, New York, Philadelphia, and Charleston flourished as coastal cities, tied together by sailing ships. In the 1800s large clipper ships could travel at speeds of 22 knots and carry shipments weighing 2,000 tons. These efficient ships produced 200 to 250 times the human energy required to operate them.[15]

Ironically, Southwest Asia, the location of the world's most accessible crude oil, was the region of the world that experimented with and developed the energy derived from windmills to replace muscle power. "There is in the world, and God alone knows it, no place where more frequent use is made of the winds," is the way in which tenth-century Arab writers described the region around Iran and Afghanistan.[16] Used almost exclusively for turning grindstones and raising water to irrigate the land, windmills spread eastward to China before arriving in Europe during the Middle Ages. Between 1500 and 1800 CE, windmills provided one-fourth of the continent's energy needs, with water and muscle carrying the largest burden.

Brought to Europe by returning Crusaders in 1200 CE, windmills along with waterwheels released some humans from the debilitating labor that broke both body and spirit. As historian of technology Lynn White put it: "The chief glory of the latter Middle Ages was not the cathedrals or its epics or its scholasticism: it was the building for the first time in history of a complex civilization which rested not on the backs of sweating slaves or coolies, but primarily on non-human labor."[17]

Associating windmills and the Netherlands is commonplace and the reasons have historical and contemporary meaning. By 1500 CE, the Netherlands for economic and societal reasons became leaders in wind power. The Dutch constructed huge windmills working together to power waterwheels to drain marshes and the lakes of the Rhine River Delta. Then, they constructed dikes to hold back the water, making available for farming approximately 40 percent more land that was previously 10 feet below sea level. Lewis Mumford wrote about this extraordinary technological achievement by saying: "the gain in energy through using wind and water power was not merely direct. The mammoth machines made possible the restoration and cultivation of rich soil, reversing a historic degradation ... The windmill added absolutely to the amount of energy available by helping to thrown open these rich lands."[18]

Figure 9.2   Kinderdijk windmills, Netherlands. *Source*: From Wikimedia Commons, the free media repository. Author: Lucas Hirschegger, 26 December 2004.

The economic benefits of windmills extended beyond farming, as Dutch entrepreneurs recognized their manufacturing benefits. With the invention of the printing press, the demand for paper skyrocketed and the ability of windmills to power presses to turn wood pulp into paper created an economic boom. In the process, of course, Europe's already depleted forests suffered another setback. By 1900, nine thousand windmills provided 90 percent of the country's energy needs, while an estimated 100,000 windmills were located throughout Europe, reducing the continent's dependence on human and animal muscle even further.[19]

In the United States, wind and water power dominated the energy needs of producers and consumers. Although the quantities used remain approximations due to uneven record keeping, the figures are astonishing. In 1850, wind power accounted for 1.4 billion horsepower hours of work, with waterwheels providing 0.9 billion hours. Steam power derived from the burning of fossil coal was a distant third, providing 0.4 billion.[20] Although the wind power used by the sailing ships that dominated the sea lines were a part of this calculation, the findings verify the important role that wind and water power played in the nineteenth-century economy and society. For the remainder of the nineteenth century, "harbors still contained forests of sailing vessel masts and the dominant feature of many rural landscapes was the windmill."[21] However, in the post Civil War era, the United States had already begun replacing renewable energy sources with non-renewable fossil coal and oil. By 1890, wood, water, and wind represented only 10 percent of the nation's energy needs. An all-encompassing paleotechnic age with its predictable fuel source had arrived.

The Paleotechnic World: Energy from Coal

Coal became the combustible energy that fueled the industrial world of the nineteenth century. Like other fossil fuels (e.g., peat, oil, and natural gas), coal was the product of millions of years of accumulated dead plant matter compressed by the pressure from above and from thermal heat within the Earth. As Alfred Crosby noted, "It is the reduced and compacted remains of forests millions upon millions of years old, forests of club mosses, some thirty meters high, of giant horsetail rushes, of flora in a profusion barely equaled by our most luxuriant jungles today."[22]

As described in Chapter 6, mining coal for commercial and household use predated industrialization. By 1600, mining was an established industry in England, Germany, Belgium, and France, with England's output reaching a million tons annually. By 1700, London consumed 500,000 tons a year, with use rising with each passing decade. The legendary London fog was the combination of the city's rainy weather and its belching residential and manufacturing smokestacks. The adoption of the "hard energy path," with coal becoming the predominant fuel source, resulted from the convergence of discoveries of plentiful seams of coal, the depletion of forests making charcoal prohibitively expensive, and the demands of an expanding iron industry. The invention of coke as fuel for the metal industry, the use of the steam engine to power machinery, and the growing network of railroads transformed towns, cities, and the landscape.

Industrialization was, in part, a steam engine revolution that changed the balance of the world's economic power. The textiles produced in Britain's mills by the end of eighteenth century destroyed India's ancient industry, displacing thousands of workers. During that century, 70 percent of the world's gross domestic product (GDP) was divided in almost equal parts between India, China, and Europe. By the end of the nineteenth century, China's share was reduced to 7 percent and India's to 2 percent, with Europe controlling a massive 60 percent and with 20 percent for the United States. A colossal shift in world power was unfolding, caused mostly by a new energy flow unknown before in human history.

The global reallocations of power changed the world's demography. Steam powered the world's factories and its transportation system of ships and railcars. The external movement of people, begun in earnest during the Age of Discovery in the fifteenth century, was catapulted ahead with the invention of the steamship and the replacement of the multi-sailed clipper ships. Four hundred thousand of these ships left Europe each year for overseas colonies and the Americas from 1850 to 1900 and then one million a year from 1900 to the outbreak of the Great War in 1914. During the same period, millions of Indians and Chinese faced with poor economic prospects migrated to the Americas, South and East Africa, and the Pacific Islands. A new energy regime based on fossil coal allowed 100 million people to move from their country of origin to a new homeland between 1830 and 1914.[23] Railroads provided for the internal movement of people from farms and towns to cities and from established settlements to frontier locations. The paleotechnic age had begun in earnest.

The use of coal for producing steam in the United States lagged behind Europe and China where the depletion of woodlands for charcoal production had taken place centuries before. America's abundant forests with their accessible and cheap wood competed with coal well into the nineteenth century. In river transportation, the use of wood or coal depended on the cost of moving the fuel from its point source to the furnace. Once the calculation of costs was stabilized and it became clear that about one-half ton of coal could replace about two tons of wood at half the cost, the transition to coal occurred rapidly.

Anthracite (hard) coal competed with wood for household heating in the early decades of the nineteenth century because of the proximity of the mines in eastern Pennsylvania to major coastal cities that were accessible by rivers and canals to the coast. The discovery of the great Pittsburgh seam, however, a large, horizontal, thick band of bituminous (soft) coal located along the rivers, transformed the economy into a fossil coal society.[24]

The adoption of less costly bituminous coal for generating steam power and as a fuel in the production of iron and steel made it the industry's principal energy source. By mid-century, the nation's mines produced only 8.4 million tons of coal but thereafter the figures point in the direction of an expanded coal-mining industry and accelerated industrialization and urbanization. In 1870, 40 million tons were mined; 30 years later, 270 million tons and at the end of World War I, an all-time high of 680 million tons. With fluctuations caused by the Great Depression (1929–38), World War II (1941–5), and the postwar economic expansion of the 1950s, coal production stabilized at 500 tons each year.[25]

The industrial use for coal grew with the introduction of the Bessemer process for making steel rails for a growing transportation network. By 1880, 5.2 million tons of coal was used to produce 3.3 million tons of coke to make 1.4 million tons of rolled steel, half of it for rails. During that decade railroads owned 90,000 miles of track. The increased number of large steam-driven locomotives consumed much larger quantities of fossil coal than the amount used in the production of steel. With the railroad transportation revolution came more railroad mileage, more track and greater horsepower, from 21.8 million in 1880 to 48.5 million by 1900. By 1885, a milestone had been reached as coal consumption eclipsed the consumption of wood for all purposes.

In Europe, coal achieved primacy much earlier in its industrialization process, since prior centuries of deforestation had removed this resource from the stock of available fuels.[26] In China, the use of coal was limited by slow industrialization and the equally slow development of a national railroad network. By 1933, only 6,200 miles of track had been built carrying 45,000 passengers in 19 of its provinces.[27] Interrupted by the Japanese invasion in the 1930s, World War II, and the civil war that followed, fossil fuel usage for power production would await the Communist takeover of 1949 and its centralized plan for massive industrialization.

Coal arrived in cities on railcars and barges and was stored in large piles before distribution to power plants and factories. Later in time, they would perform the same function for electrical power plants.

As David Nye has pointed out, "The coal wagon was a common site, hauling energy to homes and small businesses. Coal was also transformed into gas to light the city streets, first in Baltimore in 1816 and soon in other major centers."[28] By 1828, New York City's Broadway was lit by gas flares and by the 1840s coal gasification had spread to many mid-sized northeastern cities, including Hartford, New Haven, Newark, and Reading, Pennsylvania. By 1860, 183 gas companies operated in American cities and after the Civil War, with the manufacture and laying of iron pipes, coal gas was used for household cooking.

The urban atmosphere in the eighteenth- and nineteenth-century cities became thick with sulfuric smoke, ash, and tar that corroded the facades of buildings and ate away at porous statuary. Not until advances in medical science was the damage to human health recognized as respiratory diseases, tuberculosis, and pneumonia became commonplace in these contaminated environments. New sources of energy also brought with them new dangers. The mining industry experienced explosive growth to support a new steam-powered economy. By 1912, 750,000 miners labored below ground in this dangerous extraction industry, blasting, digging, and transporting about one billion metric tons annually.

Fossilized energy buried for millions of years was released in a century. Twenty-five percent of the workforce was made up of boys earning subsistence wages under degrading work conditions. Mine disasters became a common occurrence – fires, explosions, asphyxiation, flooding, and collapse – and received attention from newspapers but little mine safety enforcement from government. With the creation of the federal Bureau of Mines in 1910, a comparative study of mine death rates in Germany, France, and Britain pointed out that deaths in the United States were double those occurring in Germany and three times greater than those in France and Britain.[29]

Once railcars delivered coal to cities, burning it for industrial and residential use contaminated the air and gasification delivered a poisonous, noxious, and potentially explosive substance into America's households. As iron piping entered the design and construction of households, this inexpensive fuel made heating and cooking with coal gas a convenience that outweighed the hazards for American consumers. Despite poorer air quality, energy from coal represented an improvement in the quality of life in much the same way as electricity, water

Figure 9.3 Breaker boys at a North Ashland colliery. *Source*: From Wikimedia Commons, the free media repository. This file is from the UMBC Photographers & Collections Index (http://aol.lib.umbc.edu/specoll/photoglist.php). Accession no.:73-02-295.

and sewer pipes, telephones, and "labor-saving devices" would add to a consumer's list of desires that would soon become necessities.

In the industrialized world today, coal and its soot, ash, and tar no longer contaminate urban air or degrade the urban infrastructure of buildings, bridges, and roads. In the newly industrializing world of South and East Asia, however, coal combustion in urban areas is reminiscent of nineteenth- and early twentieth-century European and North American cities with high levels of air pollution. As noted earlier, Americans mined 1 billion metric tons of coal in 1900, the amount rose to 3.5 billion metric tons in 1950 and in 2000 reached the astounding figure of 5.2 billion. Electric power plants, located hundreds or even thousands of miles from the end-users, burn millions of tons of coal daily, mined in distant ecosystems to satisfy our growing demands for electricity, a "clean" energy. End-users are twice removed from point sources, the mines and the coal-fired power plants.

Our demand for electricity has risen five times since 1950. Then, Americans consumed 2,000 kWh of electrical energy per person each year but today the amount is 32,700 per person.[30] Now, we have coal-fired power plants emitting sulfur dioxide that combines with ozone in the upper atmosphere and is carried by the prevailing winds to distant locations where it comes to earth as acid rain leaching calcium from the soil and nutrients from the water, killing plant and aquatic life. Ecosystems with fossilized energy are exploited, the areas severely impacted by deep and surface mining, while other ecosystems receive the benefits of this energy without acknowledging the environmental costs, either to others or to themselves. This engine of energy consumption leaves a broad ecological footprint, yet removes modern humans from any sense of culpability in the transaction.

Despite these upward trends in the use of coal to produce energy today, coal's dominance as an energy source for all sorts of power ended in the 1930s, replaced by oil and natural gas. By 1960, oil and natural gas accounted for 73 percent of fossil fuel use and by the end of the century slightly less at 72.8 percent, with coal providing 22 percent and renewable sources (water, wind, and biomass) and nuclear power making up the remainder. The persistent demand for electricity, however, forced utilities to double their power production capacity every decade from 1912 to 1972 and by that time coal had reasserted itself, representing 58 percent, as the major energy source. By the end of the century, Americans consumed 36 percent of the energy for their homes and 26 percent was used for transportation. By contrast, industry consumed 38 percent of the energy but much of that was produced to satisfy Americans' demand for automobiles, appliances, frozen foods, and entertainment.[31]

"The United States has consumed more energy per capita than any society in human history, and it remains the world's leading energy consumer."[32] Today, most North Americans (living in Canada and the United States) use twice as much energy as Europeans with no appreciable difference in standard of living, 10 times more than the average Latin American and more than 100 times a typical African, with the distance between North Americans and all others getting greater except for the emerging economies of Asia (China, India, South Korea, and Singapore). Although Americans have benefited greatly from nature's bounty in supplying them with ample woodlands, natural water flows, and fossil fuels, it was in the twentieth century that the United States became the highest energy consumer in the world and the discovery of oil and the technology to refine it put the country on the hard path of oil dependency.

The Neotechnic World: Energy from Oil

What is oil? It is an ancient substance created about 50 million years ago in areas of the world that were submerged below the surface of the Earth in warm waters wonderfully suited for microscopic life to flourish and die and to repeat this process many millions of times. As such, the floor of the water became an organic graveyard that hardened into a thick layer of nutrient-rich rock. As nearby rivers discharged sediment into these waters, it was buried under enormous pressure and solidified into sandstone five miles thick. The weight from above and the thermal temperatures from below pressure-cooked the nutrient-rich fossilized microscopic life. In the process of cooking, the biological molecules of hydrogen and carbon were transformed into a complex hydrocarbon brew that we know as petroleum. Unlike coal, where the main ingredients in the transformation are plants, in the brew that becomes petroleum the important ingredients are microscopic.

Microscopic life contains more hydrogen that is more easily transformed into a hydrocarbon than coal. Petroleum is a blend of liquid hydrocarbons and gaseous hydrocarbons. When processed and refined the former become kerosene, gasoline, and semi-solid asphalt while the latter become propane, butane, and methane or natural gas. In this gaseous state, the hydrocarbons exist as billions of bubbles adding fluidity to the already liquid petroleum. Unlike coal, however, which will remain in the ground until it is mined, liquid petroleum with this gaseous pressure-cooked characteristic will press up against the miles of sandstone seeking release from the underground. Over the many millions of years, much of the oil reached the surface and spilled onto the ground as the methane leaked away. Unaccountable trillions of barrels of oil and natural gas have gone this way over thousands of years.[33]

To describe the world's eventual dependence on oil the economic historians Sam Schurr and Bruce Netschert have divided the oil age into three somewhat arbitrary periods. They titled these periods according

Figure 9.4 Oil platform in the North Sea. *Source*: From Wikimedia Commons, the free media repository.

to the purposes for which petroleum was used as energy for illuminants, as fuel oil, and as fuel for internal combustion engines. The history about the discovery of oil by drilling a well in Titusville, Pennsylvania on August 27, 1859 is well known. That discovery ushered in the age of "liquid gold" that spread from Pennsylvania, Ohio, and Indiana in the 1880s to Texas, Oklahoma, and California in the twentieth century. Drilling for oil in the oceans and in Mexico, Venezuela, and Southwest Asia in the last century marked the growing dependence on foreign oil to satisfy our "thirst" for petroleum.

Through much of our history, humans illuminated the darkness with fire, burning wood, biomass, and dried dung. As whale hunters took to the seas in the 1700 and 1800s, whale oil became a major illuminant. Its life history proved to be a short one, however, as whale hunting became a particularly destructive practice with no awareness of or concern for the finite nature of whales. As the whale populations plummeted, the price of whale oil rose to more than $2.50 a gallon and outside the reach of most users. By the 1840s, a cheap distilled petroleum product called camphene became the mostly widely used lamp oil. Despite its explosive and noxious qualities, it burned brightly.

Further experimentation with distillates led to the discovery and the production of a safer, longer-lasting and brighter illuminant known to the oil industry as "rock oil" but recognized by consumers by its trade name, kerosene. For the second half of the nineteenth century, kerosene became the leading illuminant and as "range oil" became a fuel for stoves. Ranges for cooking began to replace the cast iron stove that used wood for fuel. With the nation's population at 50 million, enough kerosene was produced annually to provide every individual with an average of 8.4 lamplight hours each day of the year. In 1880, American refineries produced 11 million barrels of kerosene with almost two-thirds exported to Europe, with Britain and Germany receiving most in the form of kerosene and lubricants.[34]

Kerosene imports provided these European customers with 125 lamplight hours for each person annually. With the further expansion of the railroads and factory systems, the need for effective lubricants for machinery to deter unnecessary wear and breakdowns became an industry priority. Once mineral oils proved their efficiency, they replaced animal and vegetable oils as industry standards in North America and in Europe. Again, both Britain and Germany, with rapidly growing transportation systems and industrial capacity, received half of the processed lubricants refined in the United States. Kerosene as an illuminant, however, began to face competition from manufactured and natural gas, especially in the growing urban markets and from the new electric power industry using Thomas Edison's commercial incandescent lighting system in the 1880s. Without finding other growing markets for its products, the future of the petroleum industry looked bleak.

Industrial fuel oil and transportation fuel oil came to the rescue. As early as the 1860s, ships had begun experimenting with oil for steam generation as the replacement for coal, because of oil's greater power, storage capacity, and as a cleaner and more efficient fuel. Locomotives as well as industrial plants did the same. Almost half of the 91 million barrels of crude oil extracted from wells in 1909 was refined into fuel oil. By this time, California had passed Texas and Oklahoma in production and was supplying western railroads with a fuel source cheaper than coal. Four-fifths of its production went to refined fuel oil in these early years. At the same time, industrial use was accelerating to

19.7 million barrels annually. Before World War I, the oil fields were producing 300 million barrels with about half consumed as fuel oil. With drilling and pumping operations using more fuel, and with an expanding export market for lubricants and even illuminants, fuel oils began to replace coal by 1920.[35]

For the first half of the twentieth century, fuel oil consumption grew without interruption, with industrial and home heating oil representing the greatest growth areas. Between 1940 and 1955, they accounted for four-fifths of the production capacity, with residential use tripling. Fuel oil also made inroads into the productive capacities of utilities to generate electricity, an increase of about 170 percent, with utilities expanding rapidly to meet the growing demand from consumers. Sales of fuel oil to commercial enterprises and residential customers doubled during this period, while those to industry and mining operations increased two and one half times. Only railroad use declined, reflecting the overall decline in the industry in face of excessive competition from truck transportation and the consumer's infatuation with the automobile.

For the first time, the United States began to import oil from foreign operators to the tune of 120 million barrels from the Greater Antilles in the Caribbean and from Venezuela. Together, these imports account for 28 percent of the refined output. Without the inclusion of oil into the energy mix, the nation would have needed an additional 200 tons of annual coal production. With coal mining increasing only by 40 million tons annually, oil production came to the rescue.[36]

The automobile, with its internal combustion fuel engine, has become the ubiquitous form of transportation in the modern world. As history has shown, however, its eventual primacy was not inevitable, because steam, electric, and internal combustion cars competed with each other in the beginning. That beginning commenced with racing competition. On June 10, 1907, eleven men in five cars left the French embassy in Beijing, China on a transcontinental journey to Paris, France. Two months later, Prince Scipione Borghese arrived first in the French capital. It was a fitting beginning for modern performance racing, using petroleum as its energy source.

The invention of gasoline as an energy source for the internal combustion engine did not eliminate the competing electric and steam vehicles until the early years of the twentieth century. Unit costs, the production of a vehicle's many components and parts, and its assembly each posed unique challenges for inventors, manufacturers, and dealerships in Europe and North America. In addition, the inertia and reluctance of consumers to accept a new mode of transportation required new and seductive sales campaigns. A major barrier to change, however, was the absence of an infrastructure that could accommodate gasoline-powered vehicles. Despite the expansion of automobile manufacturing capacity by 1910, 75 percent of the streets of major cities remained either unpaved or gravel covered. The percentage was higher in rural areas and cultural conditions help to explain the reluctance to change. Cobblestones, wood blocks, and bricks provided surface material for the few major roadways.

David Nye has explained it as a subtle weave of human interactions in the "walking city." Casual encounters took place as street vendors, delivery boys, and shoppers spoke to each other. The corner drugstore, the local saloon and café, the neighborhood store, and the front porch became the focal points for conversations. The streets became playgrounds for children and the location of social and religious events. "Residents regarded streets as sources of light and air. Because the trolley and the automobile challenged the existing street life, both were resisted; they made travel the road's preeminent function, and they were noisy and dangerous. Since by tradition people whose property abutted on a street paid for its paving and upkeep, popular resistance was manifested in the choice of uneven surfaces ill-suited for speed."[37]

Americans owned 8,000 automobiles in 1900 and most were steam-powered cars. Cars whose energy came from the internal combustion engine represented a small fraction of a small market. Although the transition from cold water to hot to steam took time, the Stanley Steamer, the most popular car, overcame this problem by heating small quantities at a time. For fuel, it used kerosene, available at every hardware store. With a reliable energy source and a simple operating system that drivers understood, it compared favorably to the internal combustion engine that required skilled mechanics and knowledge of the whereabouts of a limited number of filling stations. Electric cars never posed a serious threat to the others because heavy batteries needed repeated recharges in an environment without easily accessible electrical outlets.

The gasoline-powered engine won out over its competition because its fuel released more energy per volume of weight. In addition, the car was lighter than its competitors, so it traveled longer and faster on a tank of gasoline and got stuck less often in the rutted, muddy, seemingly impassable roads than the heavier, underpowered steamers. In 1900, a car cost on average $1,000 or the equivalent of two years' wages for the average working person in the United States. At such a prohibitive cost, the vehicle was a luxury for the wealthy only. Within a decade, however, mass marketing and mass production of a reliable and cheap car carried the day.

Marketing automobiles to consumers has a long and seductive history. For many decades, the new advertising profession invented planned obsolescence, with yearly changes in style, color, size, and optional equipment to sell cars. As decades passed, emphasizing a marketing strategy rather than automobile performance would undermine the American industry and allow foreign manufacturers to capture an increasingly large share of annual sales. As professionals, namely doctors, middle management, entertainers, and others purchased expensive cars, advertisers used high-profile individuals as role models for the general population to emulate. Financing luxury purchases that led to growing consumer debt remained in the distant future. For the present, however, the only way to emulate was to purchase a car that one could afford.

Henry Ford, the founder of Ford Motor Company, paved the way for a cheaper and reliable car for the mass market. Before he began to revolutionize transportation, Americans owned few passenger cars in the late nineteenth century. In 1908, the Ford Model T rolled off the assembly lines in Detroit, Michigan and sold for $850. Selling 6,000 Model Ts meant it remained beyond the purchasing power of most Americans. However, Ford captured the imagination of Americans by making minor stylistic changes and cutting the price each year. Just eight years later, the Model T sold for $360 to 600,000 buyers and by 1923 for under $300, without compromising performance, as a basic no-frills vehicle. Americans embraced the gasoline-powered car enthusiastically and it soon became, along with home ownership, a quintessential example of status in an increasingly status-conscious consumer society.

By the end of the 1920s, Americans owned 78 percent of the world's automobiles. One in five Americans owned a car, while in Britain and France it was one in 30 and in Germany, one in 102. As the principal purchase for most Americans, a number of industries supporting its manufacture, its maintenance, and its energy supply were adapted, expanded, or created by this ubiquitous form of transportation. By the end of the decade, an already robust steel industry in rails, girders, and angles had committed 20 percent of its production to plate steel for cars and trucks. The plate glass industry expanded rapidly to fill orders to auto manufactures that represented 75 percent of its total production in plate glass. Long before the invention of synthetic rubber for tires, tire makers used 80 percent of the world's natural rubber supply, with imports coming mostly from Asia.[38]

Ironically, the internal combustion engine that Americans now love was seen as a way to clean American cities of the manure and urine voided daily by thousands of horses on its city streets. At the moment at which Henry Ford's Model T was about to transform the world of mechanized transportation, complaints about the urban horse reached a crescendo. In New York City in 1908, with its 120,000 horses, one authority wrote about the urban horse in the following way: "an economic burden, an affront to cleanliness, and a terrible tax upon human life."[39] Such complaints coincided with growing urbanization, for as far back as 1300 CE Europeans had noted the relationship between the number of horses and its filthy streets. On average, an urban horse produced about 20 lb. of manure and gallons of urine each day and with 3–3.5 million urban horses in 1900, the sanitation and public health burden became staggering. About 17 million additional horses provided muscle power for agriculture, rural transportation, and recreation for the majority of Americans still living in small towns and villages.

As urban and environmental historian Joel Tarr has pointed out, "The evidence of the horse was everywhere – in the piles of manure that littered the streets, attracting swarms of flies and creating an offensive stench; in the iron rings and hitching posts sunk into the pavements for fastening horses' reins; and in the numerous livery stables that gave off a mingled smell of horse urine and manure, harness oil and hay."[40] As a long list of ancillary industries have matured to service the modern automobile culture, the eotechnic world of 325,000 horses in the cities of New York and Brooklyn, with a combined human population of 1,764,168, was supported by 427 blacksmiths, 249 carriage and wagon makers, 262 wheelwright shops, and 290 leather emporiums making saddles, bridges, stirrups, and harnesses.[41]

As cities became mechanized with electric trolleys, motorcars, and trucks, urban horses began to decline precipitously even though the number of horses nationally remained high at 20,091,000 in 1920. With their eventual demise, sanitation and public health agencies would point to the cleaner streets, the steep decline in breeding flies, improved air quality with the end of airborne manure particulate matter, and the faster-moving traffic of people and product. As the internal combustion engine eliminated one set of urban environmental problems, however, unbeknown to its promoters and boosters, one day its fatalities would include not only victims of collisions but also those dying from its pollutants. According to estimates developed by the World Health Organization (WHO), air pollution kills approximately 133,000 people living in industrialized cities each year. In the United States, the estimate is placed at 30,000 deaths caused by car emissions.

By 2007, the United States consumed 20.7 million barrels of oil each day. This amount constituted 25 percent of all the oil used globally each day for manufacturing and transportation in cars, trucks, and airplanes. Each gallon of gasoline burned emits 20 pounds of carbon dioxide (CO_2) or a total of 1.5 billion tons each year into the atmosphere. Countries following the United States in the order of their usage included China, Japan, Russia, Germany, and India. Combined, they used less than the United States. Without national and international restrictions on usage, one estimate puts American consumption at 32 million barrels a day by 2020.[42] As noted earlier, the pattern began with the rise of the automobile as the primary mode of transportation and the slow but steady decline in rail and public transportation. Its impact on the density and the spatial arrangements of the population placed the country on a path of dependency on oil.

In 1910, the majority of Americans lived in small, densely populated cities where fewer than 1 percent of the population owned a car. By 2007, there were 130 million cars on its streets, roads, and highways, or, read another way, 752 cars for every 1,000 people in the United States. While no other country can match this ratio, the number of cars in China, 11 cars for every 1,000 people, has been doubling every five years for the past 30 years and in the mega-city of Shanghai the number of cars and trucks is projected to quadruple by 2020.[43]

The path to dependency for the United States was not the path followed by every country, despite the recent fossil-fuel-dependent initiatives of China. For the United States, however, its oil consumption more than tripled during the 30 years from 1926 to 1955, with residential and commercial heating and gasoline showing the largest gains while manufacturing and mining uses remained unchanged. Electric power and gas power plant use more than doubled as Americans adopted the modern kitchen with its increasing array of appliances and as central heating and cooling made the transition from luxuries to necessities in the minds of most members of the middle class. Modern office buildings added to the dependency on electricity with their own enhanced versions of sealed glass towers, centrally cooled in summer and heated in winter. The need to pave more urban roads to accommodate motor vehicles only added descriptive meaning to the environmentalist's term "urban heat island" as it applied to cities. Only 10 miles of paved roads existed in 1900, while today there are four million miles of paved streets, state-maintained roads, and federal highways.

The rise in internal combustion motor vehicle ownership was accompanied obviously by a growth in gasoline use. As noted earlier, refining a 42-gallon barrel of crude oil in 1900 reflected current energy needs and at the time gasoline represented a lower priority than either kerosene for illumination or distillates for lubrication. By 1930, however, gasoline became the most important refinery product and remains unchallenged to this day. The demand multiplied six times between 1925 and 1955, from 224 million barrels to 1.5 billion barrels.[44]

The mechanization of farm equipment and its increasing use of petroleum-based pesticides and herbicides to increase agricultural productivity and the rapid expansion of the aviation industry along with the automobile during this period help to explain the dramatic growth of motor fuel usage. Passenger car manufacturing, curtailed during World War II, advanced without further interruption in the postwar period, despite the loss of the nation's energy independence. In 1946, the nation was consuming more oil than it could produce domestically and became for the first time dependent on foreign sources of crude oil. At the same time passenger car ownership almost doubled in the decade after the war (1945–55), from 25 million to 48 million automobiles.

The modern economy of the twentieth century became an energy-path-dependent oil economy whose momentum was driven by the invention of the automobile, the falling prices of cars relative to the growth in incomes of potential owners, and a quickened suburbanization in search of cheaper land for residential and commercial construction. Overall road improvements by state transportation agencies and the post World War II boom in federal highway construction, ostensibly for reasons of national security, encouraged this last development. As mega-cities expanded in the developing world, with Mexico City, São Paulo, New Delhi, Calcutta, Bangkok, and others leading the way, the phenomenon of declining population density became a trend in many of the world's cities. Although flight from decaying neighborhoods may provide a plausible explanation for this phenomenon, the impact of technological improvements in transportation, rising incomes, and the availability of cheap oil accelerated the search for modern, spacious, and affordable housing. Suburban sprawl has become synonymous with the availability of abandoned farmland on the periphery of cities but its reach extends far beyond our national borders. It is a trend that has taken hold across the developed world and its origins began in the nineteenth century with improvements in mass transportation, including horse-drawn omnibuses and trolleys, railroads, electric streetcars and buses, and eventually diesel- and gasoline-powered buses. However, the rise in the cost of public transportation and the disinvestment by state transit authorities paved the way for the passenger car.

Reliance on automobiles as the primary mode of transportation places enormous stress on the human and natural world. Before the development of safeguards to protect workers, the soil, the wildlife, and the water supply in areas near oil derricks and platforms, explosions and fires killed and injured workers and spillage and leakage contaminated the oil sites. Offshore drilling is littered with examples of explosions that sent millions of gallons of oil into the sea, killing flora, sea mammals and birds, and fish life. Once nations became dependent upon foreign sources of oil in the middle decades of the twentieth century, human error, storms, or combinations of both caused some ocean-going, single-hulled oil tankers to fracture, spilling their cargo into the deep. An ecological crisis followed for life in the seas and for those living along the shore. The energy released by ocean storms during an era of global warming may continue to make transport across the oceans by super-sized tankers a hazardous undertaking. The requirement of a double hull for new oil tankers may reduce the number of accidents.

Since most of the world's supply of non-renewable oil becomes gasoline, the refining process, despite decades of technological improvement, remains highly toxic. The air quality in areas near oil refineries is a concoction of hazardous gases and particulate matter. Elevated cardiovascular diseases and cancer rates afflict populations living in proximity to oil refineries in the modern world. Public awareness about the hazardous quality of these local environments in democratic countries restricts the construction of new refineries. Maintenance shutdowns and disruptions caused by natural disasters such as hurricanes, cyclones, and tornadoes result in fuel shortages and significant spikes in consumer prices. Instead of achieving freedom to travel and independence to live as close to or as far away from work, friends, cultural and recreational venues as one wishes, a dependence on foreign oil adds a level of vulnerability and insecurity to modern life.

Once passenger vehicles and their occupants leave the filling station, their environmental footprint becomes apparent. Banned in the United States, leaded gasoline powers motor vehicles throughout the developing world and its toxic residue is found in the blood of persons living in these urban environments. Before the ban in the United States in the 1970s, blood lead levels averaged almost 15 micrograms per deciliter of blood. By the end of the century, the average had fallen to less than two micrograms. Blood lead levels at 10 micrograms per deciliter have been associated with a greater decrease in IQ than a same-size decrease in IQ above 10 micrograms.[45] Across the globe, leaded and unleaded automobile emissions remain the largest contributor to ground-level ozone, that substance which causes haze, smog, and respiratory problems such as coughing, sneezing, shortness of breath, and asthma attacks. Measures of urban air quality caused by polluting motor vehicles in the developing world point to its continuing deterioration. In the United States, where accurate measures do exist, clean air legislation defined acceptable levels of air pollution and the reauthorization of these laws often becomes a tug-of-war among motor vehicle manufacturers, legislators, and citizens' groups about the proper emission levels based on new scientific findings about the health effects of airborne pollutants. Current conditions in the United States exceed global standards but note that poor quality air has not been outlawed. More than half of the people live in cities that do not consistently meet the federal standard and more than 80 million live in cities that consistently fail the standard.[46]

In much of the developing world, however, air quality standards remain lower on the scale of priorities than clean drinking water to limit the spread of infectious diseases, housing, health care, and personal safety. For those on the threshold of modernity, the success of the world's richest economies, namely the United States, Japan, Germany, and South Korea, serve as models for them to emulate. High economic productivity, either through factory production and/or information-related industries, leads to higher personal incomes. Higher incomes generate higher consumer spending on electrical appliances, entertainment systems, and cars. This pattern, found in advanced energy economies, is the primary reason why on a per capita basis Americans consume 7,500 gallons of oil each year and a typical Chinese uses 800 gallons.[47]

The Developing World's Demands for Energy

The pattern that developed in the industrialized world with rising economic growth and an increased energy use may repeat itself as developing countries strive to modernize. As their population growth slows and they avoid the "dirtiest" phases of air and water pollution through available technologies, they remain intent on following a path of increasing energy use. As personal incomes rise, the new middle class of India, China, Thailand, and other South Asian countries spend their disposable income on more energy-consuming appliances. Only 7 percent of the Chinese owned refrigerators before 1985; today the figure is 75 percent. The percentage owning televisions has risen from 17 to 86 during the same period. Ownership of air conditioners has risen by a factor of 50 as the demand for residential electricity more than quadrupled in the 12-year period from 1984 to 1996. To meet this growing demand, China will build 56 power plants producing electricity every year for the next decade, using its abundance of domestic coal as fuel. Without the use of modern technology to capture much of the carbon, nitrogen, and sulfur pollutants, these plants will add millions of tons of contaminants to the atmosphere. As car ownership grows by 30 percent a year in India and Thailand and quadruples in South Korea, gasoline consumption tripled between 1987 and 1997 and will double again by 2020.[48]

As the world's share of the energy economy shifts toward China, India, and Southeast Asia, as many as 300 and quite possibly 400 million new middle-class members will use their higher incomes to purchase energy-dependent consumer goods. To satisfy demand, almost half of the world's oil supply will become directed toward these growing economies. The developing world used 25 million barrels a day in 2003 and rapid growth predicts that consumption may reach 67 million barrels in 2020. During the present decade, upward revisions of yearly forecasts suggest that oil usage exceeded expectations repeatedly and that rapid growth in the automobile sector and in commercial and residential use of electricity was the driving force.[49]

In China, 11 persons in every thousand own a car, a number that looks similar to ownership in the United States almost a century ago. However, one-third of the urban Chinese possess a driver's license and almost three-quarters plan to purchase a car within five years. By these estimates, as many as 100 to 200 million cars will be on the roads of China by 2020. This projection represents many fewer cars than Americans own today but they will contribute to China's existing urban air pollution that claims about four million deaths annually. Despite government efforts to control emissions and the growth in greenhouse gases, the 3 percent of the global total currently produced by motorists is predicted to represent more than one-sixth of the global total in 20 years.

The Case for Natural Gas: A Neotechnic Energy Solution

Proximity to its source permitted Fredonia, New York, to light several buildings with natural gas as early as the 1820s. Towns near natural gas wells, such as Muncie, Indiana, and cities like Pittsburgh, Pennsylvania, experienced the availability of a cheap energy source without the visible nuisance posed by manufactured coal gas. As oil field suppliers built gas pipelines to end-users in the 1870s, natural gas passed manufactured coal gas in sales to industry. The use of cheap coal for household heating and cooking would continue into the middle of the twentieth century and would not be surpassed by oil until the 1950s and by natural gas until the oil shocks of the 1970s when the Organization of Petroleum Exporting Countries (OPEC) raised the price of crude oil to the consuming countries.

Natural gas has been promoted as a transition fuel into a post-petroleum fuel world. With a lower carbon and higher hydrogen content than coal or oil, it emits fewer pollutants and contributes less to climate change. Many energy analysts believe that natural gas, once regarded as a waste product allowed to flare off at oil field wellheads, will replace coal and oil by 2025 as the world's dominant energy source for power generation. They cite as evidence the massive investments in exploration and infrastructure, including extensive gas pipelines, taking place in diverse parts of the world, many of them major oil producers, including Algeria, Nigeria, Iran, Qatar, Turkmenistan, Russian Siberia, India, Trinidad, and Brazil.

With more than half of the world's known gas reserves located in Southwest and Central Asia, and Russia, there is enough fuel to generate power for developed and developing countries. Just as oil was a more versatile fuel than coal, offering more energy per volume and as a liquid easily transportable, natural gas meets these criteria as well. It burns hotter and cleaner than either coal or oil and it can be converted to liquid fuels and compete with oil to power vehicles. Its history as a fuel source suffered from three major liabilities, however. First, its reputation for periodic life-threatening explosions was somewhat misguided since many of the gas lights in late nineteenth-century Europe and the United States used manufactured coal gas, not natural gas. Second, gas pipeline technology failed to solve the problem of leakage. Third, oil and coal remained the fuels of choice because of their abundance and the various ways of moving them to their destinations by rail, road, barrel, and barge. Improvements in costly pipeline technology, liquefication that provided an alternative mode of transportation, global price increases in oil, and clean air legislation in industrial and industrializing countries now give a comparative advantage to natural gas.

To add to natural gas's competitive advantage beyond its cleaner high-volume methane, it contained small amounts of valuable natural gas liquids (NGLs), namely ethane, butane, and propane. Ethane is the most valuable because it can be manufactured into plastics and synthetic rubber. In addition, some gas fields produce a condensate, a gas-based liquid that can be refined into gasoline and other valuable products. As a bridge fuel into a more climate-friendly energy world, methane (CH_4) produces 50 percent less CO_2 than coal and 33 percent less than oil. Although methane is a greenhouse gas, it would reduce carbon emissions globally by 30 percent, if all coal-fired electrical plants were converted to gas-fired plants. So, a change would not eliminate the climate-changing impact of a fossil fuel energy regime but it might provide the time we need to develop alternative sources of energy.[50]

The Case for Nuclear Energy: Another Neotechnic Solution

Nuclear power plants generate electricity by bringing uranium fuel rods together to create an atomic chain reaction that produces enormous amounts of heat, turning water into steam. The steam is passed on to turbines connected to dynamos to make electricity. The energy produces no CO_2 emissions, adds nothing to the world's greenhouse gas budget, and has the highest energy density of any known fuel source. By constructing nuclear breeder reactors to produce the highly poisonous plutonium, the fuel source is limitless. Producing electricity from nuclear power plants is cheaper than coal, oil, or natural gas. A pound of uranium will produce as much heat as 1.250 tons of coal. By 2000, thirty countries were generating power using more than 400 nuclear reactors. France, with little fossil fuel, produced 80 percent of its electricity from nuclear power plants, followed by South Korea with 40 percent, Japan with 35 percent, and the United Kingdom and the United States with 20 percent.[51] Compared to the costs of using fossil fuels in terms of human health, lung diseases, acid rain, and their contribution to global climate change, nuclear power is not only cheap, it is clean.

The claim to cheap and clean is not undisputed, however. The cost of building nuclear facilities is as much as constructing new fossil coal power plants, a figure in excess of $2 billion. The claim that nuclear power emits no carbon dioxide is true only in the sense that the chain reaction itself doesn't create emissions. Mining uranium ore and refining and concentrating it to make fissionable material produce high levels of ground-level and air pollution. In addition, each nuclear plant will produce about 1,000 tons of hazardous radioactive waste each year that will be life threatening, "for more years than have passed since the last Neanderthal walked the earth."[52] The claim that nuclear power is cheaper than its fossil fuel equivalents is based on an operating cost figure of 1.8–2.2 cents per kWh. However, this figure ignores costs for research and development, plant amortization, fuel, maintenance, personnel, decommissioning, and storage for the hazardous waste. With these costs, nuclear plants can't compete economically with fossil fuel-based utilities.[53]

After the enthusiasm for the "atoms for peace" euphoria of the 1950s when proponents of nuclear power claimed that its electricity output would be so cheap that it would be futile to meter its use, the wave of plant construction worldwide slowed after a number of accidents shut down nuclear reactors. In 1957, graphite that insulated the uranium fuel caught fire at the Windscale facility in England. In the same year, an explosion of waste at the Soviet Union's Chelyabinsk plant exposed workers and surrounding villages to radioactive material. In 1969, one

Figure 9.5 Nuclear power plant at Cattenom, France. *Source*: Permission granted by Stefan Kuhn.

of the world's most powerful reactors at Saint-Laurent in France suffered a partial meltdown. A decade later on March 28, 1979, the Three Mile Island reactor on the Susquehanna River near Middletown, Pennsylvania, overheated, did not explode, but if it had would have sent potentially lethal radioactive debris over a wide expanse in the east coast of the United States. The reputation and growth of nuclear power as an alternative energy source unraveled with the explosions at the Soviet Union's reactor facility at Chernobyl, Ukraine, in April 1986. The discharge of heat was so great that it blew off the reactor's 1,000-ton roof, releasing tons of radioactive material high into the atmosphere. The death toll from the explosion, immediate exposure, and long-term death rates from radioactive induced cancers has been estimated to be more than 500,000 people. With few exceptions, nuclear power plant construction stopped globally and some plants have been decommissioned. Despite advances in the technology of nuclear power plant construction, the safeguards established to prevent accidents, and new knowledge about the annual death rates caused by fossil fuel emissions measured globally in the millions, the fear of a future Chernobyl and its human and environmental effects stymies advocates of nuclear power generation.

The Case for Renewable Wind and Solar Power: A Return to the Eotechnic

Using the wind for power predates Mumford's paleotechnic era by hundreds if not thousands of years. The energy from the wind powered sailing ships for generations and provided the power for gristmills and water pumps long before the revolution in manufacturing and industrialization. The rotation of the Earth on its axis, the uneven heating of the Earth's atmosphere by the Sun, and the world's uneven geological shape and surface contribute to the changing velocity of the world's winds. The strongest winds occur in mountain passes, along long-unobstructed continental pathways, as cold air from the Poles seeks the warmth of the tropics, and along the coastal regions where heat from the continent interacts with cool ocean waters. As a result, countries with the best coastal wind resources are numerous, with Denmark, the Netherlands, India, Argentina, and China among the most suitable to exploit this truly free energy source.

Worldwide, wind power is the fastest-growing renewable energy source. With the world's electrical current-generating capacity at 3,000 gigawatts (GW), wind power produces 40 GW. With Germany and Spain becoming world leaders in wind-turbine technology and installation, the costs of electrical power are declining rapidly and by 2010 may cost as little as 3.5 cents/kWh. Germany now generates more than 14,350 MW a year from wind power. With newly designed blades rotating at variable speeds to reduce noise and avoid bird kills, and newer turbines functioning in a wider range of wind conditions, the "energy returned for energy invested" is greatest for wind power. However, replacing dwindling oil and natural gas reserves with the power of the wind would require the installation of millions of advanced turbines worldwide, more than five times the current global manufacturing capacity. In the United States alone, about 500,000 advanced turbines would be needed by 2030 to overcome the energy losses from declining oil and natural gas supplies. In addition, the world's wind could not

Figure 9.6 Wind farm in La Muela, Zaragosa, Spain. *Source*: From Wikimedia Commons, the free media repository.

become a substitute for the burgeoning global transportation culture and agricultural infrastructure.[54]

The Sun is the ultimate source of virtually all of the Earth's energy and finding ways to collect and use it to generate electricity is an ongoing scientific and technological enterprise. Depending on the positioning of a home and its location on Earth, a typical household receives some form of passive solar heating. In North America, it amounts to about 22 watts per square foot (200 watts per square meter). Knowledge about the Sun's heating capabilities must have attracted early humans whose genes controlling pigmentation mutated to protect them from its dangerous ultraviolet rays. Ancient Greeks and Chinese used glass to focus its energy to start fires. Modern thermal electrical generators operate on a similar principle by heating water to produce steam to turn such generators. The photovoltaic effect, first discovered in 1839 by Edmund Becquerel, a young French physicist, uses direct sunlight to produce an electrical charge. In the 1950s, the first silicon solar-electric cells were used to collect the Sun's energy. Major advances in cell technology have occurred as the space program needed solar cells to power orbiting satellites.

Despite these advances, less than 1 GW of photovoltaic generating capacity exists globally. Its expense seems to be a major barrier since storing the electrical energy in a bank of batteries is cumbersome and costly. As costs decline with technological innovations, industrial capacity, networked transportation systems, individual motor vehicles, and households may become their own individual electrical utilities with the capacity to connect to a larger electrical grid as needed. As production costs for solar collectors, a controller, a transformer to change current from direct current (DC) to alternating current (AC), and the bank of batteries decline, the incentive for individuals and for governments to convert to solar energy becomes an attractive alternative to heat-trapping gases.

The demand for oil will continue unabated into the next decades as industrializing countries choose growth over conservation. Searching for the better life that includes food, housing, education, and medical care will take precedence over imposing limits. Fossil fuels now supply most of the world's energy needs. A century ago, about 100 million barrels of oil were produced annually but today the figure exceeds 20 billion barrels. If supplies of this non-renewable fossil fuel decline as predicted, steep price increases will undoubtedly dampen demand and more technological efficiencies will be built into the system.

Energy use has increased an estimated 20 times since 1850, however, and almost five times since 1950. In that year, the per capita use of electrical energy in the United States was 2,000 kWh but leaped to 32,700 per person in 2000. Factor in the revolution in consumer electronics, with televisions and computers leading the way, and the larger electric household with kitchen appliances, central heating and cooling as major contributors to kilowatt usage, and you have a near-complete reason for this staggering increase during a half-century. Coal-fired plants producing electricity for households across great distances insulate individuals from the ecological impacts of burning 5.2 billions of tons annually in 2000.

Many consumers view electricity as clean energy at the end of the pipe, not acknowledging that the source in Mumford's terms is a

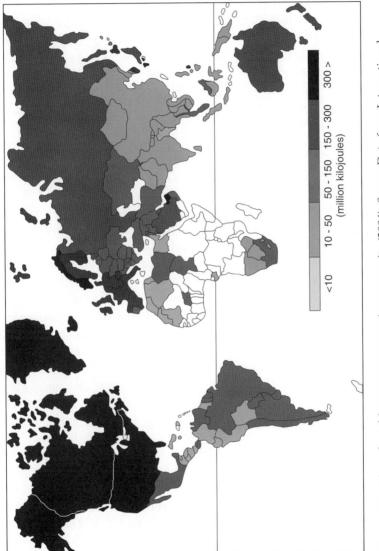

<10	10 - 50	50 - 150	150 - 300	300 >

(million kilojoules)

Figure 9.7 Map of world energy consumption per capita (2004). *Source*: Data from International Energy Agency.

paleotechnic "dirty" fossil fuel. Combined with energy from oil, the pollution from coal contributes to a host of respiratory diseases, including asthma, bronchitis, emphysema, and lung cancer. Unlike oil, known coal reserves will probably last for generations and technological efficiencies will continue to find ways to maximize its energy content and capture its pollutants. Already under way in the developed world, that effort faces daunting challenges. For example, China, with its large reserves of coal, plans to build 562 new coal-fired electric plants by 2020 to meet its growing demand. Although natural gas challenged coal globally for electric generation in the 1980s, its price increases have put coal once again in first place. At 52.2 percent, coal is now the fuel of choice for electricity generation. Although the plans to build 154 new coal-fired plants in 38 of the United States by 2030 may seem small compared to China's projections, remember that the former has a population of 300 million compared to the latter's 1.2 billion. As higher prices for non-renewable fossil fuels promote conservation and energy efficiencies, alternatives, in the form of natural gas, biomass, wind and solar power, nuclear energy, and hydrogen, require research and development support from private industry and government.

Fossil fuel energy consumption causes most of the world's air pollution and contributes greatly to water pollution and to changes in the chemical composition of the soil. The burning of coal and oil releases tons of soot, tar, and other particulate matter as well as carbon, sulfur, and nitrogen oxides. In the frozen masses of both Poles, airborne soot deposits from global fossil fuel combustion accelerate snowmelt. The darkened snow absorbs heat from the Sun rather than reflecting it back into space as light. Water pollution from acid mine drainage and mountain-top removal to expose coal deposits, oil spills on the oceans and in proximity to oil fields, and refineries themselves remain an environmental hazard. Major landscape transformations are brought about by networks of highways and suburban sprawl, strip mining of pristine landscapes, reservoirs created by damming wild and scenic rivers, and by the high-voltage electric transmission corridors and storage facilities for coal, oil, and liquefied natural gas.

Fossil fuels used globally in iron, steel, and plastics production and the chemical syntheses of pesticides and herbicides to increase agricultural productivity have intensified in recent decades. The release of millions of tons of carbon dioxide rose about 40 percent in the past 150 years. Converting grasslands and forests to agriculture and to commercial and residential development contributed to this rise as the world's capacity to transform CO_2 into oxygen declined. Along with other greenhouse gases, including methane and nitrous oxide, atmospheric changes have contributed to rising surface temperatures. The complexity of these interactions and their effects on global climate change will be the focus of the next chapter.[55]

CHAPTER TEN

A WARMING CLIMATE

Introduction

The final collapse of the last Ice Age (96,000–11,600 BP) marked the beginning of the current global warming cycle. Rapid atmospheric warming, causing abrupt rises in global sea levels, ushered in the Holocene,[1] possibly the most significant climate event in the past 40,000 years. It eliminated the thousand-mile-wide frozen Bering Straits used by human hunters to cross from Siberia to North America. Possibly beginning as early as 32,000 BP but no later than 11,000 BP, scientists believe that successive waves of hunters populated the Americas. Following the retreating ice much earlier on the Eurasian continent, from 33,000 to 26,000 BP, migrating humans from Africa and Southwest Asia replaced the *Neanderthals* in Europe.

Abrupt climate changes punctuated the transition to the warmer, more familiar modern world. Some of these changes were one-third to one-half as great as the transition from ice-age climate conditions to modern warming, roughly a change of 11–13 °F (6–7 °C).[2] Fluctuations from warm, temperate, and interglacial to cold, arctic, and glacial occurred within centuries. The climate in some regions shifted quickly, without much warning, in a few years. Evidence from deep-sea sediments and from ice cores refutes the widely held belief that the Holocene warming remained stable over time. To the contrary, the data show that the warming was interrupted by a series of abrupt cold periods, with some lasting for centuries.

Within 40 years after the beginning of the Holocene, temperatures reached modern levels. Other changes were equally dramatic. Within a decade, the warmer world led to a threefold drop in wind-blown sea salt, a sevenfold drop in wind-blown dust, and a rise in Greenland

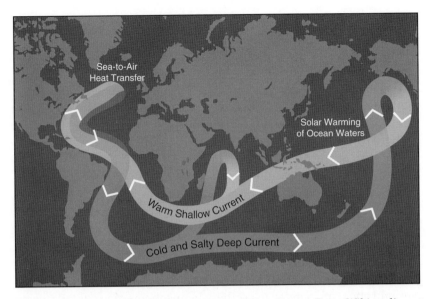

Figure 10.1 The Atlantic Deep Water Circulation. *Source*: From Wikimedia Commons, the free media repository. US Global Change Research Program (www.usgcrp.gov); www.usgcrp.gov/usgcrp/Library/nationalassessment/LargerImages/SectorGraphics/Coastal/belt.jpg.

temperatures of 9–18 °F (5–10 °C). With the creation of more wetlands, atmospheric methane (CH_4), a greenhouse gas, rose worldwide.[3]

Warming and glacial melt at about 9500 BP flooded the Black Sea basin. The biblical "flood" may have been a reference to this natural catastrophe. In the first 6,000 years of the Holocene, wetter conditions dominated the summer climates of northern Africa, India, and Southwest Asia, with frequent monsoon events. This "African Humid Period," strengthened by hotter summers and the long-term variations in the Earth's orbit around the Sun, changed the ecology of the Sahara. Northern Africa became a well-watered fertile plain with woodlands, plants, and lakes inhabited by crocodiles and hippopotami. The transitions from wet to dry occurred within decades, not millennia as previously thought, and the mechanism that may explain the movement in and out of these climate regimes is found in the Atlantic Deep Water Circulation, graphically represented in the Figure 10.1 and described more fully later in this chapter.

The release of large volumes of freshwater into the Arctic and North Atlantic Oceans by the melting of glacial ice caused by early Holocene warming at 8,200 BP disrupted the Atlantic Deep Water Circulation. The collapse of glacial Lake Agassiz, draining about 100 trillion cubic meters of freshwater into Hudson's Bay in less than one year, was such a disruptive event causing a climatic reversal.[4] Before the abrupt cooling, temperatures in Greenland were slightly higher than they are today. Of a shorter duration, global climate conditions 8000 BP may have been colder than at any time since the Younger Dryas, 12,800 BP.

Farming populations, the descendants of Indo-Europeans who in progressive migrations had brought farming technology into western and northern Europe from Southwest Asia many thousands of years before, were forced by the cold climate to retreat to warmer regions along the Mediterranean and southeastward to Ukraine, to Southwest Asia, India, and into northwest China.

Changes in the Earth's orbit along with changes in ocean circulations brought about an increase in northern hemisphere sea ice, an expansion of the Greenland Ice Sheet, cooling temperatures in Europe, and drought in North and South America.[5] In West Africa, lower amounts of rainfall caused a decline in the size of the forest's canopy. Reduced groundcover with a mature but compromised root structure retained less groundwater. Exposure to the Sun accelerated evaporation and within decades the ecology of northern Africa began to change into its current desert condition.

The once-fertile pastoral civilization of North Africa's Sahara collapsed, forcing the migration of its inhabitants to the Nile River Valley about 7,500 BP. This settlement along the Nile coincided with the millennia-long rise of the ancient Egyptian civilization. Between 7000 and 6500 BP, the Egyptians established their first empire and within centuries built the Great Pyramids at Giza. The Harappan civilization in the Indus Valley flourished as well, constructing public buildings and private dwellings of mud and fired bricks and using geometric plans to organize its cities. Among the many reasons for the rise and fall of civilizations, climate instability during the Holocene warming became a primary one.[6]

A third protracted cold period brought drought to the irrigation-dependent Fertile Crescent civilizations. The Indus Valley civilization faced a similar fate. Some archeologists suggest that the lush naturally irrigated landscape in modern southern Iraq may have been the location of the biblical "Eden." Recent archeological research has verified that the collapse of the great agricultural Akkadian Empire in northern Mesopotamia in 4200 BP coincided with a major volcanic eruption and a subsequent climate shift from wet to dry that lasted for more than a century. Sediment cores from the North Atlantic, off the coast of Africa, and from the Persian Gulf revealed dust from Mesopotamia five times greater in 4200 BP than from modern sediment cores. Three hundred years later in 3900 BP, the supply of dust in the oceans decreased to modern levels. Along with a thin layer of volcanic ash, the evidence of a great drying event and drought created the conditions for the collapse of the Akkadian Empire.[7] These concurrent events forced this ancient population to leave the north and migrate into southern Mesopotamia (modern Iraq). The Old Kingdom of Egypt and numerous villages in ancient Palestine suffered a similar fate. A fourth little ice age from 2060 BP to 400 CE drove Germanic tribes through the borders of the Roman Empire and eventually led to its destruction. Northern Asiatic tribes from Mongolia pushed southward, overwhelming the Chinese Empire.

The Rise and Fall of the Mayan Civilization

A frigid northern hemisphere turned tropical and sub-tropical regions into cooler and dryer climates. In Central America, the Mayan civilization expanded its agricultural productivity northward into the Yucatan, now part of Mexico, and built pyramids and cities in areas formerly thick with tropical vegetation and malaria-bearing mosquitoes. Its thriving urban centers maintained a population density of as many as 500 persons per square mile during its zenith. Rising agricultural productivity released persons to engage in manufacturing and promoted the development of artistry.

The drying and cooling came at a cost, however, as too many years without rainfall caused a series of collapses in Mayan agricultural productivity. Sedimentary records suggest that severe droughts began 1300 CE and revisited the region for the next 500 years. The Mayans abandoned some cities as early as 1240 CE and the remaining ones by 1190 CE when another severe dry period hit the area. Long-term drought brought famine. Malnutrition led to compromised immune systems, disease, and death. Other causes may have contributed to the demise of the Mayans but the relationship of climate change to the collapse is a compelling one.[8] After this particular global cooling ended, a tropical hydrological cycle gained strength as the climate warmed. The tropical forest returned along with the mosquitoes and forced the remaining Mayans to abandon their homes and to migrate southward. The fact that Mayan ruins are discovered now in the dense tropical rainforests of Central America is evidence of a warming climate.

The Medieval Warm Period (1000–1300 CE) and Little Ice Age (1300–1850 CE)

Although many gaps still exist in our knowledge about climate warming and cooling, efforts to unravel the complexities of the global climate have focused on specific events that trigger changes in the weather. E. E. Lamb writing about climate change and history used the phrase *Medieval Warm Epoch* to describe these events and was also more precise in locating the warming geographically. The warmest conditions existed first in European Russia and Greenland between 950 and 1200 CE and later in most of Europe from 1150 to 1300 CE.[9] Because of a warming climate, the Norse cultivated large areas of Iceland and established settlements in Greenland and Labrador beginning in the tenth century. Yet, by the twelfth century, the open sea-lanes of the North Atlantic were frozen over and Greenland and Labrador settlements disappeared. Iceland's farming communities also retreated in the wake of the advancing ice. By 1300 CE conditions deteriorated further only to ameliorate somewhat from 1400 to 1500 CE before sinking into a further freeze by 1600 CE. Lamb's analyses used a number of historical sources, including tree line and vegetation changes, preliminary tree ring measurements, and Greenland ice core samples.

Recent analyses of the *Medieval Warm Epoch* by the United Nation's Intergovernmental Panel on Climate Change (IPCC, 2007), however, note the absence of conclusive data about estimates of a global medieval warm period. These data indicate that northern hemisphere mean temperatures were warmer from 950 to 1100 CE than at any time in the previous 2,000 years yet still below the warm temperatures recorded for the last two decades of the twentieth century.[10] A final conclusion from the IPCC strikes a cautionary note about both the medieval period and the present. "The climate was unlikely to have changed in the same direction, or by the same magnitude, everywhere. At some times, some regions may have experienced even warmer conditions than those that prevailed throughout the 20th century. Local climate variations can be dominated by internal climate variability, often the result of the redistribution of heat by regional climate processes."[11]

This conclusion applies to discussions about the Little Ice Age (1300–1850 CE) that engulfed much of the northern hemisphere. Again, regional climate variability needs to be taken into account when discussing the impact of the Little Ice Age on the affected populations. Overall food production plummeted in pre-industrial Europe where diets consisted mostly of bread and potatoes. Even in the best of times, food consumption seldom exceeded 2,000 calories for the majority of the population. For these most vulnerable populations, widespread malnutrition was followed by famine and the outbreak of infectious diseases. Not all populations suffered equally during this extended freeze, however. For those living along major rivers and along the coasts, fishing, and in some cases ice fishing, provided the animal protein lacking in the diets of the majority.

The bubonic plague followed the great European famine in 1400 CE. Between 1100 and 1800 CE, France experienced frequent famines, 26 in 1100 and 16 in 1800. Increasing cold temperatures shortened the growing season by at least one month in northern European countries and the elevation for growing crops retreated about 60 feet. In New England, 1815 was called the year without a summer.[12]

Historians debate the effects of the Little Ice Age on very high-profile events that potentially changed the direction of human affairs. In 1588, the remaining ships in the Spanish Armada, outmaneuvered and battered by Queen Elizabeth's nimble seafarers, sought safety in retreat by circumventing the British Isles on their way back to Spain. As they proceeded northward, the ships encountered a low-pressure system of hurricane proportions in the North Atlantic. The weakened Armada was ravaged by the storm, making another assault impossible, and solidifying Britain's maritime position.

In another instance, the Little Ice Age played a role in historical events. George Washington's Revolutionary Army at Valley Forge, Pennsylvania bore winter's harshest blows in 1777 and yet emerged as a disciplined fighting force at winter's end in 1778. They crossed the frozen Delaware River to attack and defeat a surprised mercenary army of Hessian soldiers at Trenton, New Jersey. The crossing and the victory represented a major turning point in the American Revolution.[13]

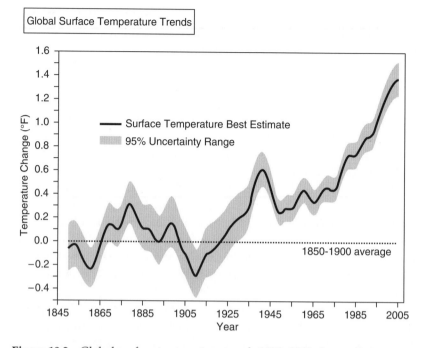

Global Surface Temperature Trends

Figure 10.2 Global surface temperature trends 1850–2005. *Source*: Data source: Brohan, P.J.J., Kennedy, I. Harris, S.F.B. Tett and P.D. Jones. 2006. Uncertainty estimates in regional and global observed temperatures changes: a new dataset from 1850. *Journal of Geophysical Research* 111;D12106. doi:10.1029/20D3JA009974. © Crown copyright 2006: data provided by the Met Office. Permission granted by Pew Center on Global Climate Change. www.pewclimate.org.

By 1812, Napoleon's armies, having achieved military success across continental Europe, found themselves bogged down in Imperial Russia, facing a rejuvenated Russian Army and a brutal early winter accompanied by sub-zero temperatures. Napoleon's retreat and the loss of his army represented the beginning of the end for his continental ambitions and the First French Empire. After 1850 and without warning, the Little Ice Age came to an end. Increased solar energy with elevated sunspot activity, the impact of industrialization on atmospheric concentrations of greenhouse gases, and changes in the Atlantic Deep Water Circulation have been identified as the probable causes either alone or in combination.

Current Global Climate Conditions

Scientists identify three basic causes for climate change: the exchange of energy by the oceans and atmosphere, fossil fuel emissions, and solar energy. With global temperatures rising since 1860, IPCC (2007) scientists predict that temperatures will continue to increase by at least 3.2 °F and as much as 7.2 °F by 2100. Evidence of global warming appears in a melting Arctic ice cap – a reduction in volume of 3 to 4 percent each decade since the 1970s, or an area the size of Texas and Arizona combined. According to the National Sea Ice Center, in 2007 Arctic sea ice melted at the rate of 81,000 square miles a day, an area as large as Kansas. Although some of the ice returns in the winter, 2006 and 2007 marked the lowest amount of accumulated Arctic ice ever recorded, as it rained well above the Arctic Circle in December and January of both years. As a result, sea levels that have been rising since 1800 have accelerated during the past half century. Predictions of global sea-level rise by as much as one foot to a foot-and-a-half fall within the probable range.

A global climate change disaster would accompany accelerated melting of the Greenland ice sheet. It would raise sea levels worldwide by 23 feet and create catastrophic changes in the lives of the 600 million people who live in coastal zones at or below sea level. Rising sea levels would submerge coastal areas and affect the natural migration of plants, animals, and microbes. Although no scientists predict a complete melting, Greenland's melting increased by 30 percent between 1979 and 2007. In 2007, the melt was 10 percent higher than in any previous year. Today, the East Greenland ice sheet adds 257 cubic kilometers of freshwater to the ocean each year from both melting and icebergs breaking off into the North Atlantic. This amounts to the total volume of almost three Chesapeake Bays and is greater than all of the ice in the European Alps. By 2100, those levels are estimated to reach 456 cubic kilometers per year. With land-based runoff factored in, the total freshwater flowing from Greenland into the Atlantic is estimated to increase from 438 cubic kilometers to 650 cubic kilometers by 2100.[14] Satellite images, measuring small changes in the Earth's gravitational field, noted that the mass of Greenland decreased by 50 cubic miles of ice in 2005.[15]

As many regions of the world became uninhabitable, however, newly exposed land caused by melting and warming temperatures would

Figure 10.3 Blomstrandbreen Glacier, Svalbard, halfway between Norway and the North Pole. The top photo was taken in 1928; the bottom one in 2002. *Source*: Permission granted by the Norwegian Polar Institute and Greenpeace.

make more land available for vegetation and habitation. During warming phases in the northern latitudes, migrating forests would replace northern tundra and formerly marginal lands may become suitable for cultivation and food production. Rising temperatures and volatile weather systems in the tropics and middle latitudes, however, would displace millions and possibly billions of people. The regional benefits for many would be outweighed by the impending disasters facing billions of people, plants, animals, insects, and microbes.

Although the specific impacts of a warming global climate system on various populations in different parts of the world remain a subject of ongoing research, climatologists point out that during the current warming phase we can expect the physical properties of CO_2 to contribute to more rainfall, higher atmospheric and oceanic temperatures, more clouds, and higher wind velocity. The biological effects of CO_2 may also contribute to longer growing seasons in the temperate climates. Former arid and semi-arid lands mainly unavailable for agriculture may receive sufficient moisture to increase food stocks for a global population whose growth rate may stabilize at nine billion people by 2050 CE. With an unequal distribution of rainfall, however, formerly well-watered environments may become stricken by long uninterrupted periods of drought. An impending catastrophe with no short-term solutions may be unfolding, as we continue to use the planet's atmosphere as a carbon sink by dumping millions of metric tons of greenhouse gases into it each year.

The Role of Solar Energy

This role draws on the research of the Russian mathematician and scientist M. M. Milankovitch, who argued that the eccentric orbit of the Earth was the reason for its major global climatic cycle every 100,000 years. During that time frame the planet goes through a full interglacial/glacial cycle. Within this longer pattern, another one, 41,000 years in duration and caused by the tilt of the Earth on its axis, controls the amount of solar energy reaching Earth's higher latitudes.

A much shorter cycle caused by the "wobble" of the Earth on its axis occurs either at 23,000- or 19,000-year intervals and affects the amount of radiation striking the low latitudes and the Equator. Milankovitch argued that during the past 800,000 years, Earth experienced eight complete glacial/interglacial cycles. Ice ages lasted for approximately 90,000 years followed by 10,000 years of warming. Accordingly, the current interglacial phase, the Holocene (Anthropocene), should be coming to an end.

Since the Milankovitch explanation accounts for only 0.1 percent change in the total solar energy reaching the Earth, however, climatologists, building on his earlier theory, have developed a more precise explanation for the driving force behind climate change. They argue that fluctuations in solar energy follow a cyclical pattern of sunspot activity. Using this pattern, they have identified a pattern of eight cycles during the past 720,000 years of Earth history. Slight variations in the Earth's orbit change the distribution of sunlight striking the Earth at different latitudes during different seasons. It is a pattern that takes thousands of years to complete, as noted by the length of these cycles. They are 90,000 years in length from full glacial reflecting up to 80 percent of the sunlight back into space. With a warming climate and meltwater replacing ice and snow, its reflective capacity is less than 0.1 percent. Actually, it's more like 0.07 percent, which means that ocean water absorbs light rather than reflecting it.

The Role of the Atlantic Circulation Energy Exchange

By studying the isotopic composition of the Greenland ice sheet, we know that the global climate has fluctuated abruptly during the past 100,000 years. Many scientific hypotheses exist about the causes for these climate shifts, yet one stands out, namely that the world's oceans serve as a heat transportation system absorbing much heat from the solar energy penetrating the atmosphere.

When ocean water freezes in the Arctic, it pushes out its salt content into the remaining ocean water. Salty water is heavier than freshwater so it sinks as freshwater from the tropics evaporates, loses heat and is pulled northward toward the Arctic. When it reaches Greenland, a massive body of heavy salt water is sinking to the bottom of the ocean, all the while attracting lighter freshwater from the tropics. A circulation pattern moving heat around the globe, often called a "conveyer belt," is established. In other words, the world contains an energy exchange mechanism that regulates the global climate system. The ratio of salt water to freshwater maintains climate equilibrium. Increase the volume of salt water to higher levels and a rapid long-term cooling commences.

Since rising temperatures increase freshwater snowmelt, they change the salinity of the oceans that affects global circulation. Currently, the

Greenland ice sheet sends river water into the Arctic sea and the melting of the Arctic ice pours freshwater into the North Atlantic Ocean. Decreasing the volume of heavy salt water disrupts the great Atlantic Deep Water Circulation that brings warm tropical water across the Equator toward the North Pole. These warm waters become the Gulf Stream that warms the New England coast and brings moisture and warmth to the British Isles. Without it, these coastal regions would become several degrees colder, turning fertile soil into permafrost. Oceanographers have documented in recent decades a decline in the Atlantic Ocean water's salt content.

Since the oceans transport heat, abrupt small changes in temperature or increases in fresh glacial melt lower the density of the water, namely its capacity to sink, slowing down and in severe instances stopping the circulation. According to some climatic models, turning off or slowing down this current, often called "a huge heat pump," has cooled down the northern temperate regions and been responsible for many of the abrupt climate oscillations during the past 100,000 years. As noted earlier, evidence from the Greenland ice cores revealed an abrupt cooling of from 2 to 6 °C (35.6–42.8 °F) 8200 BP, with cold conditions taking place in Europe and North America. Coupled with a large atmospheric decrease in methane (CH_4) concentrations, large-scale changes in atmospheric circulation from the Arctic to the tropics caused widespread drought.[16]

The more sunlight they absorb the warmer the ocean waters become, as the feedback system accelerates, contributing to faster warming of the oceans and the climate system. During the last warming interglacial period, about 125,000 years ago, sea levels were between 13 and 19 inches higher than in 2005, due mainly to glacial ice melt and average polar temperatures 3 to 5 °C higher than now because of differences in the orbit of the Earth. Assuming that the current warming follows a similar cyclical pattern, with the Earth's orbit now contributing to the warming rather than serving as the major forcing mechanism, then the fossil fuel emissions since the beginning of industrialization may result in changes that even the IPCC (2007) cannot forecast.

None of the computer models used by researchers for the IPCC (2007) predict a complete shutdown of the Atlantic Deep Water Circulation in the twenty-first century. However, the predicted range from zero to a 50 percent change in this century leaves the matter open to continued scientific investigation. Continued emissions at current levels, inducing further atmospheric warming, will result in reduced salinity and

Figure 10.4 The dry bed of the Sabarmati River in Ahmadabad, India.
Source: Associated Press/Ajit Solanki.

ultimately affect the deep ocean. An accelerated melting of the Greenland ice sheet, predicted to happen in the coming decades, has commenced. A total melting that would raise global sea levels by 23 feet and create worldwide havoc would take several centuries to complete, according the IPCC researchers. A collapse of the West Antarctic ice sheet would shorten this scenario by centuries and add another 20 feet to sea-level rise. Such a projection would trigger a cooling, suggesting the beginning of an ice age.[17]

The Role of Fossil Fuel Emissions

The IPCC released a summary of its fourth report titled *Climate Change 2007*. Working for more than six years, 2,500 scientific experts from 130 countries contributed to the report. Their findings were stunning: "Warming of the climate system is unequivocal, as is now evident from observations of increases in global average air and ocean temperatures, widespread melting of snow and ice, and rising global mean sea level. Eleven of the last 12 years (1995–2006) rank among the 12 warmest years in the instrumental record of global surface temperature."[18] The report also notes that "the atmospheric concentrations of carbon dioxide in 2005 exceed by far the natural range over the last 650,000 years (180–300 ppm) as determined from ice cores."[19]

Rising global temperatures translate into increased atmospheric water vapor, a greenhouse gas, as more of the world's warming ocean water evaporates causing more precipitation. The recent decades have been the warmest in the past 600 years. Fourteen of the 15 warmest days on record in the United States have occurred since 1980. A 2005 panel of the National Academy of Science identified 1998 as the hottest year on record, followed closely by the years 2001–4.

Others, including the American Geophysical Union, argued that natural causes such as sunspot activity or volcanic eruptions cannot explain the rapid increase in global near-surface temperatures.[20] Solar energy output has periodic and sometimes irregular patterns of high intensity occurring during 11-year cycles, with those of low intensity taking place about every 3.5 years. Variations in the Sun's energy output fail to explain, in any scientific way, rising global temperatures.

Near-surface temperatures are higher now than they have been in the past 2,000 years. If the current rate of change continues into the lifetimes of our grandchildren, the global climate will change at a rate faster than at any time since the end of the last Ice Age, 11,600 years ago. And if this trend continues to the end of the twenty-first century, global temperatures may rise to levels not seen in the past two million years.

Releasing the energy sequestered for millions of years in fossilized plants and animals by burning coal, oil, and gas elevates concentrations of carbon dioxide (CO_2) in the atmosphere. Most of these emissions come from three key sectors, electricity generation, transportation, mostly automobiles, and buildings. Electric power generation and heat together produce 42 percent of all CO_2 emissions in the United States and 41 percent worldwide. Prior to industrialization, the world's atmosphere contained CO_2 levels of 280 parts per million (ppm), up from the Pleistocene levels of 190 ppm.[21] Such low levels prohibited the cultivation of food crops. The pre-industrial levels of 280 ppm had remained stable for thousands of years, dating back to the development of the world's first cities.

With carbon dioxide remaining in the atmosphere for as long as 100 years, CO_2 levels rose slowly during the early stages of industrialization, taking 150 years to get to 315 ppm. The invention of the internal combustion engine and the refining of fossil oil into gasoline accelerated atmospheric accumulation of CO_2. The emissions from Henry Ford's first Model T (1909) continue to circulate in the atmosphere, as do the emissions from every car, power plant, and airplane built and operated since then, everywhere in the world.

The combustion of fossil fuels and deforestation have increased the CO_2 load in the atmosphere to its 2005 level of 378 ppm.[22] Within the past hundred years, 40–50 per cent of the world's pioneer forests and uninhabited lands that change CO_2 into oxygen by the process known as photosynthesis became farmland, open ranges for domestic cattle and sheep, and sites for commercial and residential development. From 1850 to 2000, the human contribution to the increased concentration of CO_2 by burning fossil fuels, deforestation, and agriculture was about 1.7 trillion tons.

The United States contributed 30 percent of those emissions, followed closely by the European Union countries, with China contributing 7 percent and India about 2 percent. About 40 percent of it remains in the atmosphere and continues to increase at a rate of about 0.5 percent per year. If current trends continue, carbon dioxide concentrations will reach 500 ppm by 2050, concentrations not seen since the Eocene, some 50 million years ago, when the Earth was a greenhouse. Such concentrations will trigger a global rise in temperature of 11 °F (6 °C), a change significantly larger than what separates us now from the last Ice Age.

The IPCC (2007) states that only a 3 percent probability exists that temperatures will reach this catastrophic level globally, triggering mass extinctions and creating a *terra incognito*, a world unknown to humans. Human life will survive but on a transformed and more desolate planet. However, some scientists argue that a 5.5 °F rise in temperatures that are within the range of IPCC (2007) predictions will create carbon cycle feedbacks by 2050, causing a surge in greenhouse gases. These scientists predict that creating a global climate much warmer than the present one will cause feedbacks that accelerate greenhouse gases within a few decades after 2050. An unstoppable escalation of CO_2 emissions, causing the loss of the permafrost or the Amazon rainforest, will take us to 800–1,000 ppm.[23]

Without these carbon feedbacks, a 5 °F rise in global temperatures creates its own disasters. Only once in the past three million years, with sea levels about 80 feet higher than present levels, has the planet been as warm. At these levels, reached over a period of three centuries, eastern seacoast cities, including Boston, New York, Philadelphia, and Washington, and most of Florida would be lost to rising sea levels. Lost land in China would displace 250 million people, in Bangladesh 120 million, and in India about 150 million.

The global pattern of a warming climate has a scientific basis that is explained clearly in the following way. Global warming occurs more rapidly in frigid regions rather than in the temperate and tropical areas because arctic air lacks water vapor. This characteristic makes CO_2 a more important greenhouse gas where the air is cold and dry. In warmer, humid air, water vapor is a more important transporter of heat than CO_2. Also caused by the burning of fossil fuels, other fast-growing greenhouse gases such as methane (CH_4) and chlorofluoro-carbons (CFCs) with greater heat-absorbing qualities than CO_2 have affected atmospheric temperatures. Much of the world's frozen surface in the uppermost latitudes of Siberia and the Arctic are permafrost zones.

Land that remains frozen for two years or more is labeled permafrost and its depth may range from a few hundred to thousands of feet. In some regions of eastern Siberia it runs up to a mile deep. The top layer of the surface supports the growth of grasses, shrubs, and large spruce trees depending on the duration of normally short summer growing seasons. Given the harshness of the Arctic and Siberian climates, however, the normal cycle of growth and decay is often interrupted by the onset of the polar winter. The result is dead but only partially decomposing frozen biomass. With the repeated cycles of thawing and freezing, partially decomposing organic material is compressed and pushed into the frozen permafrost where it can remain for millennia depending upon climate conditions.

So, much like peat and coal deposits, the permafrost becomes a sink for accumulated carbon. Rising global temperatures and the thawing of the top layer of the permafrost since the 1980s expose partially decomposed organic matter, some of which dates from the middle of the last ice age 50,000 BP. Exposure and a completion of the process of decay releases millions of tons of methane (CH_4), a more powerful greenhouse gas than CO_2.[24]

Figure 10.5 Melting permafrost in Siberia. *Source*: United States Geological Survey, http://carbon.wr.usgs.gov.

What Is to Be Done?

As of this writing, no significant action has been taken by the major emitters of greenhouse gases to slow the continued growth of fossil fuel use. Although the European Union has made significant progress in getting its members to plan for a return to 1990 levels of greenhouse gas emissions by 2020, the United States has followed a business-as-usual scenario. The European Union suppresses oil consumption by imposing fossil fuel taxes on users. California, the world's seventh largest economy, has taken a leadership position in the United States by requiring a decrease of 30 percent in automobile-created greenhouse gases by 2015. The federal Department of Energy has been unwilling to uncouple economic growth from the growth in the use of fossil fuels. In this regard, it has the support of some energy-producing companies, a few of whom have supported disinformation campaigns alleging that the science of climate change is either faulty or inconclusive. To exacerbate the current state of affairs in the United States, federal fuel economy standards for automobiles established by legislation almost 30 years ago remain stuck at 24 miles per gallon. In fact, a significant loophole in this legislation exempted "light trucks" from the standards governing cars. As a result, a generation of sports utility vehicles (SUVs) entered the United States consumer market, exempt from lax fuel economy standards and supported by cheap global oil prices. Although global oil prices are rising rapidly as the result of consumer demand in the United States and growing industrial economies in China, India, and elsewhere, no current comprehensive international energy policy has emerged since The United Nations Framework Convention on Climate Change (1992) and the Kyoto Protocols (1992, 1997), the latter being the agreement that the United States refused to endorse in 2001.

If the current business-as-usual pattern continues, greenhouse gas levels will rise significantly, ice sheets will melt at an accelerated rate, and a further warming of the global climate, in the 5–10 °F range, will probably cause some of the catastrophic changes described earlier in this chapter, including sea-level rise of as much as 80 feet over the next three centuries. An alternative scenario includes the legislation adopted by California, and would result in $150 billion annual saving in oil imports if adopted nationally. It would slow down the export of national treasure, amounting to $700 million annually, to oil producing and exporting countries (OPEC). If such a policy were followed for 35 years, it would save seven times the estimated amount of recoverable oil located in the Arctic National Wildlife Refuge, and reduce CO_2 emissions with the goal of stabilizing them at 450 ppm. Most scientists believe that this figure must be achieved in order to avoid the worst consequences of global warming.

To achieve this goal, an alternative scenario described by economist Robert Socolow as "stabilization wedges" fixes CO_2 emissions at the current level of seven gigatons (a gigaton is a billion metric tons) a year. The business-as-usual scenario will cause emissions to rise to 14 gigatons a year in 50 years. The following are examples of wedges that Professor Socolow suggests to stabilize emissions and prevent runaway climate change: Replace two billion cars that get 30 miles per gallon with two billion cars getting 60 miles per gallon. In the electricity-generating sector, use carbon sequestering and storage technology at 1,600 large coal-burning, electricity-generating power plants. Replace coal power generation with one million 2-megawatt wind turbines. Building 700 gigawatts of nuclear power will displace coal power that produces gigatons of emissions. These and many other combinations will stabilize emissions at the current seven-gigaton annual level. All of these "stabilization wedges" come with challenges and costs but they and many others provide us with alternatives to our hard fossil fuel energy path dependency that places us in the current predicament.[25]

What can be done to avoid the probability of impending disasters, the magnitude of which was described earlier, is to follow a prudent middle course by making serious cuts in greenhouse gases over the next decades. Conservation and efficiency measures could lead the way by improving fuel economies in automobiles, lighting, and the full range of household appliances and electronics. Despite the gains to be made through efficiencies, all carbon fuel users need to confront the expensive reality of burning fossil fuels. These substantial environmental costs need to be measured and passed on to all users. One small example outlines the variable costs that must be assigned to fossil fuel users. A person driving a fuel-efficient automobile that gets 60 miles to the gallon pays a considerably lower carbon tax and may even receive a tax credit, while the driver of a sports utility vehicle gets charged the full amount for doing so. So, one of the most effective measures to deal with what is becoming an urgent task is to methodically, consistently, and systematically impose revenue-neutral carbon taxes on fossil fuel users. The revenues could then be used to offset some of the costs of slowing climate change.

Carbon taxes, anathema in some modern societies convinced by their elected officials into believing that consumption of material goods bears no ecological costs, are consistent with the Kyoto Protocol. One of its defining programs encouraged developed countries, the major emitters of greenhouse gases, to slow down their rate of emissions early in this century while developing countries adopt clean technologies, rather than following the hard fossil fuel energy path dependency.

The United States and Australia's refusal to join the Protocol has resulted in the failure of others, including developing countries such as China with huge natural reserves in coal, to follow a "clean" energy path. Others, including Japan, have refused to make additional commitments to reduce emissions without contributions from the United States. Brazil's efforts to slow deforestation in the Amazon rainforest may be reinvigorated with commitments from the United States.

Even without its participation, the Kyoto Protocol (1997) has borne important results. Kyoto's "cap and trade" policy of establishing emissions targets coupled with emissions trading has accumulated 30 billion dollars in greenhouse gas trades since its inception in 2005. Targets state clearly the amount of emissions to cut and emissions trading allows heavy fossil fuel users to buy credits from low emitters at market rates. As global emissions rise, the cost to high emitters also rises, encouraging them to adopt cleaner technologies. The European Union's regional trading system and Kyoto's Clean Development Mechanism lead the way in using "cap and trade" methods effectively.

In order for the Kyoto Protocol to continue functioning, it will require reauthorization in 2012. Given recent developments in the United States with the creation of its Climate Action Partnership, a consortium of 10 major corporations, and pending congressional legislation to cut emissions by 60 to 80 percent by 2050, an alternative scenario seems probable in the near future.

EPILOGUE

While Earth history and human history are interwoven in complex and important ways, no species has changed the natural world as significantly as *Homo sapiens*. Similarly, staggering shifts in global geology, climate, and ecology created an environment conducive to both the creation of life and its extinction. Placed in this context, the history of *Homo sapiens*, relative newcomers to the planet Earth, can be thought of as either mostly potential with much of its history about to unfold or as thin as its threadlike past. Driven by complex processes at work in the atmosphere, the biosphere, and the oceans, the global climate system altered their habitats, shaped their evolutionary spiral, and directed their migration out of Africa. It fostered the growth of intelligence by forcing *Homo sapiens* to continually adapt, surmounting, for the sake of their very existence, the obstacles presented by the natural world.

The abrupt climate transition from the glacial Pleistocene to the warm, wet, and higher atmospheric CO_2 Holocene provided the switch that transformed human affairs and led to the invention of agriculture. As archeologist Charles L. Redman has pointed out, "The introduction of agriculture is regarded by many as the single most important transformation in human history. And the ensuing rise of early civilizations is among humankind's greatest achievements."[1] After millennia of hunting and gathering, the invention of agriculture imposed some form of work schedule for cutting and collecting cereals. With increases in climatic fluctuations and temperature changes, different species of wild plants matured at different times. Monitoring these different rates of maturity would require making notations, committing this new knowledge to memory, and working with other members of the group to schedule times to harvest and store wild cereals. "These processes would have invited sowing and carefully measured storage, processes that would have had to occur within a carefully monitored calendar frame."[2]

Weaving baskets for food storage in pits and in dry caves and creating an inventory of the harvest became necessary prerequisites for successful settlements, the precursors of the village. The production of new pragmatic ideas and new symbols noted changing climate conditions. New cultural representations signified group identity and cultural cohesion within an emerging village life.

The importance of agriculture in the modern world can hardly be overestimated. It stimulated settlement and sedentary life ways, the building of permanent structures for storage and living space, the acquisition and accumulation of consumer goods, and a growth in material wealth. It expanded the need to protect existing goods and their means of production. "[These] may have been a key step in the growth of social hierarchies as well as militarism and probably led to further movement of people into closely packed settlements that could be more easily defended."[3]

Overcoming its precarious beginnings, agriculture sustained the stability of the world's population over many millennia and its rapid rise during the twentieth century. Population growth accelerated with improvements in agriculture and trade. Increasing surpluses of food set the stage for an expansion in the production of goods and services. As a market-driven capitalist revolution took place in the eighteenth century, it depended on these increasing surpluses. Agriculture and industry became mutually reinforcing as commercialization and the marginalization of small-scale farmers gave way to the imperatives of the market for agricultural products. Surplus laborers flocked to the growing cities, finding work in manufacturing and industry.

The impact of extensive farming changed the landscape in ways unknown to thousands of past generations of foragers and hunters. On the negative side, ecosystems that contained a diversity of plant and wildlife became intensively cultivated monocultures. Farms replaced forest, grass, and wetlands and in the process "invited the development of disease epidemics by providing high densities of genetically uniform hosts."[4] It also changed the social relationships between the producers and consumers of food. Market economies based on agricultural surpluses released the creative energies of non-farm individuals to innovate, invent, and experiment on a large scale for the first time in human history.

With the advance of worldwide commercial farming, new mechanical technologies for plowing and harvesting, new chemical fertilizers, pesticides and herbicides, new biological hybrid seeds, and new methods of redirecting and extracting water for irrigation improved crop yields. They also disrupted natural ecosystems in ways that we are only now beginning to understand more fully. Water pollution accelerated with the massive use of pesticides and herbicides on farmlands. And the social, economic, and environmental costs of intensification to boast crop yields seldom got factored into food prices.

Agricultural expansion and the ability to sustain productivity in synchronization with explosive global population growth may be related to the destabilizing of the world's climate. Today, farmland consumes about one-third of the Earth's surface and 40 percent of the world's crops are produced by only 16 percent of the planet's irrigated land.[5] Transforming virgin soils, burning crop residues, increasing yields by using nitrogen fertilizers, and applying herbicides to destroy insects release carbon dioxide, methane, and nitrous oxide into the atmosphere.

These gases account for about 20 percent of the total human contribution to greenhouse emissions. Land clearance accounts for another 14 percent. With a significant growth in human population since 1850, some 15 percent of the world's forests have been converted to agriculture with 30 percent devoted to crops and the rest to pasture. Cultivation released as much as 20 to 40 percent of the carbon stored in these woodlands.

Agricultural production accelerated as farmers increased their use of fossil fuel energy in the form of machines, petroleum, and chemicals. In the past 30 years alone, farmland grew by more than 27 million acres while the number of farm workers continued to plummet. The retreat of the forests in much of the developing world and the loss of natural habitats and soil nutrients are the present-day costs of agriculture's bountiful harvest.

Coincidentally, fuel crises in history followed population growth. Although the spread of agriculture played a significant role in widespread deforestation, charcoal shortages created by economic and population growth stymied early manufacturing. As manufacturing capacity expanded into factory-based industrialization, fuel crises hit concentrated industries the hardest, especially consumer industries such as textiles where consumer incomes and changes in the size of the population proved to be critical to an industry's growth. Population growth slowed and consumer purchasing power waned as incomes declined in the wake of these crises.

Unlike the rapid transformation, replacement, and invention of new manufacturing processes from the nineteenth into the twenty-first century, the early manufacturing of metal took place in thousands of small enterprises throughout the countryside. The fabrication of finished iron products was divided almost equally between country and city. And since so much of iron production of tools, nails, and weapons was commissioned by the military, it was not subject to changes in consumer demand. In later centuries, economies of scale and the concentration of production in the hands of a few suppliers would signal the end of much small-scale manufacturing and the beginning of industrialization.

The organization of production and the extensive division of labor in large-scale factory operations became important symbols of industrial development, as did steam engines, steam turbines, and internal combustion engines in replacing muscular with mechanical power. The age of the industrial machine required a major increase in mineral output. Metal parts had to meet the requirements of durability, tensile strength, and the capacity to be made in shapes and forms needed by machines that translated processes into products. Unlike machines of wood, metal could be cast, riveted, and welded to operate at greater speeds, accept greater strains, become more compact, and bear more weight.

Mining, smelting, fabricating, and transporting metal machines and metal goods demanded sources of fuel with more energy density than wood. Releasing the Sun's energy sequestered in fossil coal and oil by increasingly mechanized mining and drilling operations would provide the fuel for the age of industrialization. It transformed the value of fossil fuels such as coal and petroleum into many-billions-of-dollars industries in the early decades of the twentieth century in Europe and the United States and in the early decades of the twenty-first century in China.

Technological advances accelerated industrialization and the emergence of mass consumption of many products derived from petroleum. Burning millions of tons of coal generated ever-increasing amounts of electrical energy to power up consumer capitalism's many appliances, electronic devices, and gadgets. The environmental costs of industrialization, hardly trivial during the thousands of years of manufacturing, would be measured in despoiling the land with open pit and mountaintop destruction, mine drainage, deteriorating air quality, habitat loss for other species, and microclimate changes that would become the harbingers of regional and global climate change.

At the end of the twentieth century, the industrialized world was a world of steel. The physical infrastructure of each industrial and industrializing country depended on steel. Their economies depended on mills and factories producing steel products for expanding commercial and consumer markets. Their societies came to depend on many consumer products, automobiles not the least among them, in which steel remained the material of choice for frames, enclosures, cabinets, motors, and mechanical and electrical connectors. Japan, South Korea, Germany, France, Britain, and Canada continue to increase their share of steel imports, not only for automobiles but also in concrete-reinforcing bars, plate and sheet steel, and pipe and tube steel. The share produced and consumed by China and India will continue to grow throughout the twenty-first century.

The affluence attributed to affordable consumer goods, low food prices, and plentiful oil may be coming to an end as the world's demand exceeds its current supply. The age of energy-consuming faster and larger cars and trucks on networks of connected highways had many consequences, not all of which were predicable. The rise of the movable feast in prepared foods transported long distances by motorized vehicles increased the availability of "fast" food. What some health experts and nutritionists regard as a crisis in eating habits in much of the developed world can be traced to a growing disconnection between the household dynamics of family relationships, shared meals and conversation, and the lure of mobility and freedom provided by the automobile. Low prices, high sodium content as a preservative to prevent spoilage, and "soft" drinks high in high-fructose corn syrup as a substitute for higher-cost sugar lure increasing numbers of consumers into a nutritional trap that is difficult to escape. Eating out becomes a recognizable pattern, leaving preparation to restaurants that provide more choices than households. Eating more packaged and prepared "takeout" food requires heating only, leaving the contents entirely in the hands of the preparers and processors. "Snacking" foods high in salt replace meals, while sweeteners and high-fructose soft drinks become substitutes for water. All of these changing patterns of food consumption require a more intensive use of fossil fuel energy.

Another way of understanding the synergistic relationship and interdependencies that shape human affairs is through the prism of rising prices. After decades of declining global food prices led by the Green

Revolution (1965 to 1985) in which the world's cereal prices (wheat, rice, corn, barley, and sorghum) rose from 1 billion metric tons to 1.8 billion metric tons, cereal prices dropped by 40 percent. During those 20 years, billions of people escaped from hunger, in many cases for the first time. That trend has come to an end with no reversal in sight, unless the world's hard fossil fuel energy dependent path changes. This new unpredictable turn has happened, despite a rise in the world's global rice output of 2.3 percent in 2008. With three billion people depending on rice to provide one-third of their daily calories, its price has risen steadily since 2003, rising 141 percent in 2007. Rising yields have led to rising demands and rising prices in part because government policies have created incentives to divert grains into biofuels. Consumption of these fuels grew by 20 percent in 2008 and caused a spike in food prices. The current worldwide food crisis, which affects billions of people, has been caused in part by the faulty reasoning behind subsidizing corn growers to produce ethanol to power motor vehicles. The synergy between agriculture, industry, and energy use becomes transparent when looked at in this way.

Other factors also contribute to the current crisis. As incomes rise in countries with fast-growing economies, more expensive foods, such as meat, once considered luxury foods, became staples. In China and India, consumption of meat products rises as more of its citizens achieve middle-class status. Once a staple for consumers, grains get diverted to feeding livestock. Since wheat production, unlike rice, has failed to keep pace with rising population, per capita demand for wheat products, such as bread and cereals, exceeds supply. In 2007–8, world wheat stocks dropped to levels not seen since 1947–8, while price levels are greater than those experienced in the past 25 years.

Much of the current predicament can be traced directly and indirectly to the price of fossil fuels and involves their combustion, producing carbon dioxide that accumulates in the atmosphere for decades. In the modern world, almost every human activity contributes to these emissions. Until recently, humans have viewed the atmosphere as a "commons" to which everyone has access but no one is responsible for its health and safety. In economic terms, the cost of using the atmosphere as a sink for our waste and contributing to surface warming has been zero. Like many of the transformations described throughout this book, our hard path dependency on fossil fuels may be coming to an end.

NOTES

Introduction

1 Jerry Bentley, "A New Forum for Global History," *Journal of World History* Vol. 1 (1990) iii–v.
2 Jerry Bentley, "Why Study World History?" *World History Connected* Vol. 5, Nos. 1, 3 (2007).
3 Joseph Fletcher, "Integrative History: Parallels and Interconnections in the Early Modern Period 1500–1800," *Journal of Turkish Studies* Vol. 9 (1985) 38.
4 Daniel Lord Smail, *On Deep History and the Brain* (Berkeley, CA: University of California Press, 2008).
5 J. R. McNeill, "Observations on the Nature and Culture of Environmental History," *History and Theory* (December 2003) 6.
6 Fred Spier, *The Structure of Big History: From the Big Bang Until Today* (Amsterdam: Amsterdam University Press, 1996) 19.
7 Smail, *On Deep History and the Brain*, 190–3.
8 Spier, *The Structure of Big History*, 57.
9 Stanley H. Ambrose, "Late Pleistocene Human Population Bottlenecks, Volcanic Winter, and Differentiation of Modern Humans," *Journal of Human Evolution* Vol. 34 (1998) 623–51 and cited in Smail, *Deep History*, 194.
10 Alf Hornborg, "Introduction: Environmental History as Political Ecology," in Alf Hornborg, J. R. McNeill, and John Martinez-Alier (eds.) *Rethinking Environmental History: World-System History and Global Environmental Change* (Lanham, MD: AltaMira Press, 2007) 13.
11 Christopher G. Boone and Ali Modarres, *City and Environment* (Philadelphia, PA: Temple University Press, 2006) 43.
12 Ibid., 39.
13 Ibid., 45.
14 "What's Land Got to Do with It?" A Symposium at the Lincoln Institute, November 2007. http://www.lincolninst.edu/news/atlincolnhouse.asp.

15 R. Bin Wong, *China Transformed: Historical Change And The Limits of European Experience* (Ithaca, NY: Cornell University Press, 2000) 279.

16 Harold C. Livesay, *Andrew Carnegie and the Rise of Big Business* (Glenview, IL: Scott, Foresman and Co., 1975) 126.

17 Steven Johnson, *The Ghost Map: The Story of London's Most Terrifying Epidemic and How It Changed Science, Cities, and the Modern World* (New York: Riverhead Books, 2006) 92–3.

18 Kenneth Pomeranz, *The Great Divergence: China, Europe, and the Making of the Modern World Economy* (Princeton, NJ: Princeton University Press, 2000) 117, and Johnson, *The Ghost Map*, 95.

19 J. R. McNeill, "Yellow Jack and Geopolitics: Environment, Epidemics, and the Struggles for Empire in the American Tropics, 1640–1830" in *Rethinking Environmental History: World-System History and Global Environmental Change*, 199–217.

Chapter 1

1 David Christian, *Maps of Time: An Introduction to Big History* (Berkeley, CA: University of California Press, 2003) 26.

2 Ibid., 502–3.

3 Christian, *Maps of Time*, 62.

4 Ibid., 63.

5 Thomas M. Cronin, *Principles of Paleoclimatology* (New York: Columbia University Press, 1999) 441.

6 Ibid., 442.

7 Ibid.

8 Ibid.

9 Christian, *Maps of Time*, 71.

10 Carl Sagan and Ann Druyan, *Shadows of Forgotten Ancestors* (New York: Random House, 1992) 29.

11 Ibid.

12 Ibid., 29–30.

13 W. F. Ruddiman (ed.), *Tectonic Uplift and Climate Change* (New York: Plenum Press, 1997).

14 Christian, *Maps of Time*, 70.

15 R. M. DeConto and D. Pollard, "Rapid Cenozoic Glaciation of Antarctica Induced by Declining Atmospheric CO_2," *Nature* Vol. 421, No. 6920 (2003) 245–9.

16 Tim Flannery, *The Eternal Frontier: An Ecological History of North America and Its People* (New York: Grove Press, 2001) 101.

17 Ibid., 102.

18 Ibid.
19 Ibid., 172.
20 Douglas Palmer, *Atlas of the Prehistoric World* (New York: Random House, 1999) 139.
21 William F. Ruddiman and John E. Kutzbach, "Plateau Uplift and Climatic Change" *Scientific American* Vol. 264, No. 3 (March 1991) 68.
22 Palmer, *Atlas of the Prehistoric World*, 142.
23 Ibid.
24 Ruddiman and Kutzbach, "Plateau Uplift and Climatic Change," 68.
25 Ibid.
26 Ibid., 70.
27 Ibid., 71.
28 Ibid.
29 Ibid., 72.
30 Cronin, *Principles of Paleoclimatology*, 172–3.
31 Lisa Cirbus Sloan and Eric J. Barron, "Paleogene Climatic Evolution: A Climate Model Investigation of the Influence of Continental Elevation and Sea-Surface Temperature Upon Continental Climate," in Donald R. Prothero and William A. Berggren (eds.) *Eocene–Oligocene Climate and Biotic Evolution* (Princeton, NJ: Princeton University Press, 1992) 16, 207–9.
32 Cronin, *Principles of Paleoclimatology*, 442.
33 Ibid.
34 Gabriel J. Bowen, David J. Beerling, Paul L. Koch, James C. Zachos, and Thomas Quattlebaum, "A Humid Climate State During the Palaeocene/ Eocene Thermal Maximum," *Nature* Vol. 432 (25 November 2004) 495–9.
35 J. C. Zachos et al., "A Transient Rise in Tropical Sea Surface Temperature during the Paleocene–Eocene Thermal Maximum," *Science* Vol. 302, No. 5650 (2003) 1551–4.
36 IPCC, *Climate Change 2007: The Physical Science Basis. Contribution of Working Group 1 to the Fourth Assessment Report of the Intergovernmental Panel On Climate Change* [S. Solomon, D. Qin, M. Manning, Z. Chen, M. Marquis, K. B. Averyt, M. Tignor and H. L. Miller (eds.)] (Cambridge: Cambridge University Press, 2007) 442.
37 Donald R. Prothero, *The Eocene–Oligocene Transition: Paradise Lost* (New York: Columbia University Press, 1994) 22–3.
38 Walter Sullivan, *Continents in Motion: The New Earth Debate* (New York: American Institute of Physics, 1991, 2nd edn) 164–5.
39 Andrew Sherratt, "Plate Tectonics and Imaginary Prehistories: Structure and Contingency in Agricultural Origins," in David R. Harris (ed.) *The Origins and Spread of Agriculture and Pastoralism in Eurasia* (Washington, DC: Smithsonian Institution Press, 1996) 132.

40 Peter J. Wyllie, *The Way the Earth Works: An Introduction to the New Global Geology and Its Revolutionary Development* (New York: John Wiley & Sons, 1976) 209.

41 Ibid., 210.

42 Ibid., 211.

43 Sullivan, *Continents in Motion*, 167.

44 Wyllie, *The Way the World Works*, 210–11.

45 Sullivan, *Continents in Motion*, 170–1.

46 T. M. Cronin et al., "Mid-Pliocene Deep-Sea Bottom-Water Temperatures Based on Ostracode Mg/Ca ratios," *Marine Micropaleontology* Vol. 54, Nos. 3–4 (2005) 249–61. A. M. Haywood et al., "Global Scale Paleoclimate Reconstruction of the Middle Pliocene Climate Using the UKMO GCM: Initial Results," *Global Planetary Change* Vol. 25 (2000) 239–56.

47 Ruddiman and Kutzbach, "Plateau Uplift and Climatic Change," 66.

Chapter 2

1 Leon Croizat, *Space, Time and Form: The Biological Synthesis* (Caracas: Published by the Author, 1962) 605.

2 Peter B. deMenocal, "Plio-Pleistocene African Climate," *Science* Vol. 270, No. 5233 (October 6, 1995) 53–9.

3 Clive Gamble, *Timewalkers: The Prehistory of Global Colonization* (Cambridge, MA: Harvard University Press, 1994) 75.

4 Glenn C. Conroy, *Reconstructing Human Origins: A Modern Synthesis* (New York: W. W. Norton and Co., 1997) 125.

5 Peter B. deMenocal, "Plio-Pleistocene African Climate," 53–9.

6 T. G. Bromage et al. (eds.) *African Biogeography, Climate Change, and Human Evolution* (Oxford: Oxford University Press, 2000) and Christopher Stringer and Robin McKie, *African Exodus: The Origins of Modern Humanity* (New York: Henry Holt & Co., 1996) 149–78.

7 M. H. Wolpoff and A. G. Thorne, "Modern Homo sapiens Origins: A General Theory of Hominid Evolution Involving the Fossil Evidence from East Asia," in F. H. Smith and F. Spencer (eds.) *The Origins of Modern Humans: A World Survey of the Fossil Evidence* (New York: Alan R. Liss, 1984) 411–83. By the same authors, "The Case Against Eve," *New Scientist* Vol. 22, No. 1774 (1991) 33–7.

8 R. L. Cann, M. Stoneking, and A. C. Wilson, "Mitochondrial DNA and Human Evolution," *Nature* Vol. 325 (1987) 31–6.

9 Noel T. Boaz, *Eco Homo: How the Human Being Emerged from the Cataclysmic History of the Earth* (New York: Basic Books, 1997) 96–7.

10 E. S. Vrba et al. (eds.) *Paleoclimate and Evolution with Emphasis on Human Origins* (New Haven, CT: Yale University Press, 1995). Donald C. Johanson and Maitland A. Edey, *Lucy: The Beginning of Humankind* (New York: Simon & Schuster, 1990) 328–34. J. Desmond Clark, "The Origins and Spread of Modern Humans: A Broad Perspective on the African Evidence," in Paul Mellars and Christopher Stringer (eds.) *The Human Revolution: Behavioral and Biological Perspectives on the Origins of Modern Humans* (Princeton, NJ: Princeton University Press, 1989) 565–88.

11 R. R. Ackermann and J. M. Cheverud, "Detecting Genetic Drift versus Selection in Human Evolution," *Proceedings of the National Academy of Sciences of the United States of America* Vol. 101, No. 52 (December 28, 2004) 17946–51.

12 Luigi Luca Cavalli-Sforza, *Genes, Peoples, and Languages* (New York: Farrar, Straus and Giroux, 2000) 10–11.

13 Ackerman and Cheverud, "Detecting Genetic Drift," 17948–51.

14 Katherine Milton, "Primate Diets and Gut Morphology: Implications for Hominid Evolution," in Marvin Harris and Eric Ross (eds.) *Food and Evolution: Toward a Theory of Human Food Habits* (Philadelphia, PA: Temple University Press, 1987) 105–6.

15 Randall L. Susman, "Who Made the Oldowan Tools: Fossil Evidence for Tool Behavior in Plio-Pleistocene Hominids," *Journal of Anthropological Research* Vol. 47, No. 2 (Summer 1991) 129–51.

16 Phillip V. Tobias, *Olduvai Gorge Volume 4: The Skulls, Endcasts and Teeth of Homo habilis* (Cambridge: Cambridge University Press, 1991).

17 Henry M. McHenry and Katherine Coffing, "Australopithecus to Homo: Transformations in Body and Mind," *Annual Review of Anthropology* Vol. 29 (2000) 125–46.

18 Cavalli-Sforza, *Genes, Peoples, and Languages*, 165.

19 Richard F. Kay, Matt Cartmill, and Michelle Balow, "The Hypoglossal Canal and the Origin of Human Vocal Behavior," *Proceedings of the National Academy of Sciences of the United States of America* Vol. 95, No. 9 (April 1998) 5417–19.

20 Misia Landau, *Narratives of Human Evolution* (New Haven, CT: Yale University Press, 1991).

21 R. Bonnefille et al., "High-Resolution Vegetation and Climate Change Associated with Pliocene Australopithecus afarensis," *Proceedings of the National Academy of Sciences of the United States of America*, Vol. 101, No. 33 (August 17, 2004) 12125–9.

22 Michael Balter, "Fossil Tangles Roots of Human Family Tree," *Science* Vol. 291, No. 5512 (March 2001) 2289–91.

23 G. Philip Rightmire, *The Evolution of Homo erectus: Comparative Anatomical Studies of an Extinct Human Species* (New York: Cambridge University Press, 1991).

24 Timothy D. Weaver and Charles C. Roseman, "New Developments in the Genetic Evidence for Modern Human Origins," *Evolutionary Anthropology*, Vol. 17 (February 22, 2008) 69–80. Gary Stix, "Traces of a Distant Past," *Scientific American* (July 2008) 56–63.

25 Kate Wong, "Global Positioning: New Fossils Revise the Time When Humans Colonized the Earth," *Scientific American* (August 2000) 23.

26 Thomas J. Crowley and Gerald R. North, "Abrupt Climate Change and Extinction Events in Earth History," *Science*, Vol. 240, No. 4855 (May 1988) 996–1002.

27 Christopher B. Ruff, "Climate and Body Shape in Human Evolution," *Journal of Human Evolution* Vol. 21 (1991) 81–105.

28 Ralph M. Rowlett, "Fire Use," *Science* Vol. 284, No. 5415 (April 1999) 741.

29 Bing Su et al., "Y-Chromosome Evidence for a Northward Migration of Modern Humans into Eastern Asia during the Last Ice Age," *American Journal of Human Genetics* 65 (December 1999) 1718–24.

30 Ann Gibbons, "Ancient Island Tools Suggest Homo erectus was a Seafarer," *Science* Vol. 279, No. 5357 (March 1998) 1635–7.

31 Elizabeth Culotta, Andrew Sugden, and Brooks Hanson, "Humans on the Move," *Science* Vol. 291, No. 5507 (March 2001) 1721.

32 J. M. Bermúdez de Castro et al., "A Hominid from the Lower Pleistocene of Atapuerca, Spain: Possible Ancestor to Neanderthals and Modern Humans," *Science* Vol. 276, No. 5317 (May 1997) 1392–5.

33 Michael Balter, "In Search of the First Europeans," *Science* Vol. 291, No. 5507 (March 2001) 1724.

34 Katerina Harvati et al., "Neanderthal Taxonomy Reconsidered: Implications of 3D Primate Models of Intra- and Interspecific Differences," *Proceedings of the National Academy of Sciences of the United States of America* Vol. 101, No. 5 (February 2004) 1147–52.

35 Daniel E. Lieberman et al. "The Evolution and Development of Cranial Form in Homo sapiens," *Proceedings of the National Academy of Sciences of the United States of America* Vol. 99, No. 3 (February 2002) 1134–9.

36 Ezra Zubrow, "The Demographic Modelling of Neanderthal Extinction," in Paul Mellars and Chris Stringer (eds.) *The Human Revolution: Behavioural and Biological Perspectives on the Origin of Modern Humans* (Princeton, NJ: Princeton University Press, 1989) 212–31.

37 O. Soffer, J. M. Adovasio, and D. C. Hyland, "The 'Venus' Figurines: Textiles, Basketry, Gender, and Status in the Upper Paleolithic," *Current Anthropology* Vol. 41, No. 4 (August 2000) 511–37.

38 Ezra Zubrow, "The Demographic Modelling of Neanderthal Extinction," 217.

39 Mark Derr, "Of Tubers, Fire and Human Evolution," *The New York Times* (January 16, 2001) D 3.

40 Michael Balter, "Did Homo erectus Tame Fire First?" *Science* Vol. 268, No. 5217 (June 1995) 1570. Bernice Wuethrich, "Geological Analysis Damps Ancient Chinese Fires," *Science* Vol. 281, No. 5374 (July 1998) 165–6.

41 Derr, "Of Tubers, Fire and Human Evolution," D 3.

42 Norman Owen-Smith, "Pleistocene Extinctions: The Pivotal Role of Megaherbivores," *Paleobiology* Vol. 13, No. 3 (Summer 1987) 351–62.

43 Ibid.

44 Rachel Caspari, Sang-Hee Lee, and Ward H. Goodenough, "Older Age Becomes Common Late in Human Evolution," *Proceedings of the National Academy of Sciences of the United States of America* Vol. 101, No. 30 (July 2004) 10895–900.

45 Ehud Weiss et al., "The Broad Spectrum Revisited: Evidence from Plant Remains," *Proceedings of the National Academy of Sciences of the United States of America* Vol. 101, No. 26 (June 2004) 9551–5.

46 Peter J. Richerson, Robert Boyd, and Robert L. Bettinger, "Was Agriculture Impossible during the Pleistocene but Mandatory during the Holocene? A Climate Change Hypothesis," *American Antiquity* Vol. 66, No. 3. (July 2001) 387–411.

47 Anan Raymond, "Experiments in the Function and Performance of the Weighted Atlatl," *World Archaeology* Vol. 18, No. 2, Weaponry and Warfare (October 1986) 153–77.

48 Mark Nathan Cohen, "Prehistoric Patterns of Hunger," in Lucile F. Newman et al. (eds.) *Hunger in History: Food Shortage, Poverty, and Deprivation* (Cambridge, MA: Blackwell, 1990) 57–8.

49 Ibid., 65–7.

Chapter 3

1 Edouard Bard, Frauke Rostek, Jean-Louis Turon, and Sandra Gendreau, "Hydrological Impact of Heinrich Events in the Subtropical Northeast Atlantic," *Science* Vol. 289 No. 5483 (August 2000) 1321–4.

2 Robert B. Marks, *The Origins of the Modern World: A Global and Ecological Narrative* (New York: Rowman & Littlefield, 2002) 39.

3 Michael Rosenberg, "Cheating at Musical Chairs: Territoriality and Sedentism in as Evolutionary Context," *Current Anthropology* Vol. 39 No. 5 (December 1998) 653–81.

4 A. M. T. Moore and G. C. Hillman, "The Pleistocene to Holocene Transition and Human Economy in Southwest Asia: The Impact of the Younger Dryas," *American Antiquity* Vol. 57 No. 3 (1992) 491.

5   Bruce D. Smith, "Prehistoric Plant Husbandry in Eastern North America," in C. Wesley Cowan and Patty Jo Watson (eds.) *The Origins of Agriculture: An International Perspective* (Washington, DC: Smithsonian Institution Press, 1992) 221.

6   Andrew Sherratt, "Plate Tectonics and Imaginary Prehistories; Structure and Contingency in Agricultural Origins," in David R. Harris (ed.) *Origins and Spread of Agriculture and Pastoralism in Eurasia* (London: UCL Press, 1996) 137.

7   Steven Mithen, *After The Ice: A Global Human History* (Cambridge, MA: Harvard University Press, 2004) 23–4.

8   Robley Matthews, Douglas Anderson, Robert S. Chen, and Thompson Webb, "Global Climate and the Origins of Agriculture," in Lucile Newman et al. (eds.) *Hunger in History: Food Shortage, Poverty and Deprivation* (Oxford: Blackwell, 1990) 41.

9   Emily McClung De Tapia, "The Origins of Agriculture in Mesoamerica and Central America," in C. Wesley Cowan and Patty Jo Watson (eds.) *The Origins of Agriculture: An International Perspective* (Washington, DC: Smithsonian Institution Press, 1992) 156.

10 T. Douglas Price and Gary M. Feinman, *Images of The Past* (Mountain View, CA: Mayfield Publishing Company, 1997) 217.

11 Bruce D. Smith, "Prehistoric Plant Husbandry in Eastern North America," in C. Wesley Cowan and Patty Jo Watson (eds.) *The Origins of Agriculture: An International Perspective*, 111.

12 Mithen, *After The Ice*, 203.

13 Ibid.

14 John Noble Wilford, "An Early Heartland of Agriculture Is Found in New Guinea," *The New York Times* (June 24, 2003) D2.

15 Mark A. Blumer, "Ecology, Evolutionary Theory and Agricultural Origins," in David R. Harris (ed.) *The Origins and Spread of Agriculture and Pastoralism in Eurasia*, 40.

16 Ibid., 41.

17 Ibid., 48.

18 Wen-ming Yan, "Origins of Agriculture and Animal Husbandry in China," in C. Melvin Aikens and Song Nai Rhee (eds.) *Pacific Northeast Asia in Prehistory: Hunter-Fisher-Gatherers, Farmers, and Sociopolitical Elites* (Pullman: Washington State University Press, 1992) 114.

19 T. Douglas Price and Gary M. Feinman, *Images of the Past*, 220.

20 C. Wesley Cowan and Patty Jo Watson (eds.) *The Origins of Agriculture: An International Perspective*, 144.

21 Robley Matthews et al., "Global Climate and the Origins of Agriculture," 42–3.

22 Jean Gimpel, *The Medieval Machine: The Industrial Revolution in the Middle Ages* (London: Pimlico, 1988) 33.

23 Carlo M. Cipolla, *Before The Industrial Revolution: European Society and Economy, 1000–1700* (New York: W.W. Norton, & Co., 1976) 29.

24 Robert W. Fogel, "New Findings on Secular Trends in Nutrition and Mortality: Some Implications for Population Theory," in Mark Rosenzweig and Oded Stark (eds.) *Handbook of Population and Family Economics* (Amsterdam: Elsevier, 1997) 433–81. John Duffy, *The Sanitarians: A History of American Public Health* (Urbana, IL: University of Illinois Press, 1990). Martin V. Melosi, *The Sanitary City: Urban Infrastructure from Colonial Times to the Present* (Baltimore, MD: The Johns Hopkins University Press, 1999).

25 Myron P. Gutmann, *Toward A Modern Economy: Early Industry in Europe, 1500–1800* (Philadelphia, PA: Temple University Press, 1988).

26 Alfred W. Crosby, Jr., *The Columbian Exchange: Biological and Cultural Consequences of 1492* (Westport, CT: Greenwood Press, 1972) 184.

27 Ibid., 178–9.

28 Ibid., 186–7.

29 Marks, *The Origins of the Modern World*, 96.

30 Ibid., 103.

31 George Grantham, "Agricultural Supply during the Industrial Revolution: French Evidence and European Implications," *The Journal of Economic History* Vol. 49 No. 1 (March, 1989) 43–72.

32 Crosby, *The Columbian Exchange*, 166.

33 Ibid., 201.

34 Vaclav Smil, *Enriching the Earth: Fritz Haber, Carl Bosch, and the Transformation of World Food Production* (Cambridge, MA: MIT Press, 2001) 199.

35 J. R. McNeill, *Something New Under The Sun: An Environmental History of the Twentieth-Century World* (New York: W.W. Norton & Co., 2000) 216.

36 Ibid., 221.

37 Ibid., 223.

38 Smil, *Enriching the Earth*, 204.

39 Jason McKenney, "Artificial Fertility: The Environmental Costs of Industrial Fertilizers," in Andrew Kembrell (ed.) *The Fatal Harvest Reader: The Tragedy of Industrial Agriculture* (Washington, DC: Island Press, 2002) 122–3.

40 Ibid., 127.

41 Smil, *Enriching the Earth*, 199–209.

42 John McHale, "Global Ecology: Toward the Planetary Society," in G. Bell and J. Tyrwhitt (eds.) *Human Identity in the Urban Environment* (Harmondsworth: Penguin, 1972) 133.

43 Smil, *Enriching the Earth*, 245.

Chapter 4

1 Massimo Livi-Bacci, *A Concise History of World Population* (Cambridge, MA: Blackwell, 3rd edn, 2001) 6, 25–6. Livi-Bacci cautions that, "data on world demographic growth are largely based on conjectives and inferences drawn from non-quantitative information." (p. 25).
2 Les Groube, "The Impact of Diseases upon the Emergence of Agriculture," in David R. Harris (ed.) *The Origins and Spread of Agriculture and Pastoralism in Eurasia* (Washington, DC: Smithsonian Institution Press, 1996) 101–2.
3 Alfred W. Crosby, Jr., *The Columbian Exchange: Biological and Cultural Consequences of 1492* (Westport, CT: Greenwood Press, 1972) 30.
4 Massimo Livi-Bacci, *A Concise History of World Population*, 25–6.
5 Philip M. Hauser, *World Population and Development: Challenges and Prospects* (Syracuse, NY: Syracuse University Press, 1979) 3.
6 http://www.census.gov/ipc/www/worldhis.html.
7 Ben J. Wattenberg, *The Birth Dearth: What Happens When People in Free Countries Don't Have Enough Children* (New York: Pharos Books, 1989).
8 Mark Nathan Cohen, *Health and the Rise of Civilization* (New Haven, CT: Yale University Press, 1989) 33.
9 Ibid., 32.
10 Ibid., 112.
11 Ibid., 117.
12 Ibid.
13 Ibid., 120–1.
14 David R. Harris, "Settling Down: An Evolutionary Model for the Transformation of Mobile Bands into Sedentary Communities", in J. Friedman and M. Rowlands (eds.) *The Evolution of Social Systems* (Pittsburgh, PA: University of Pittsburgh Press, 1978) 409.
15 William H. McNeill, "The Conservation of Catastrophe," *The New York Review of Books* (December 20, 2001) 86.
16 Luigi, Luca Cavalli-Sforza, "The Spread of Agriculture and Nomadic Pastoralism: Insights from Genetics, Linguistics and Archaeology," in David R. Harris (ed.) *The Origins and Spread of Agriculture and Pastoralism in Eurasia*, 52.
17 Massimo Livi-Bacci, *A Concise History of World Population*, 25–6.
18 Ibid., 26.
19 Mary Jackes, David Lubell, and Christopher Meiklejohn, "Healthy but Mortal: Human Biology and First Farmers of Western Europe," *Antiquity* Vol. 71 (1997) 653.
20 Cavalli-Sforza, "The Spread of Agriculture ...," 230.
21 Ibid., 39.

22 Ibid., 47.

23 Groube, "The Impact of Disease ...," 125.

24 Ezekiel J. Emanuel, "Preventing the Next SARS," *The New York Times* (May 12, 2003), A25.

25 William M. Denevan, "The Pristine Myth: The Landscape of the Americas in 1492," in Karl W. Butzer (ed.) *The Americas Before and After Columbus: Current Geographical Research* (Annals of the Association of American Geographers, 1992) 82(3): 369–85.

26 Ibid., 39.

27 Robert Fogel, "The Relevance of Malthus for the Study of Mortality Today: Long-Run Influences on Health, Mortality, Labour Force Participation, and Population Growth," in Kerstin Lindahl-Kiessling and Hans Landberg (eds.) *Population, Economic Development, and the Environment* (New York: Oxford University Press, 1994) 241–51.

28 Ibid., 108.

29 Patrick R. Galloway, "Long-Term Fluctuations in Climate and Population in the Preindustrial Era," *Population and Development Review*, Vol. 12, No. 1 (March 1986) 7–14.

30 Geoffrey Parker, *Europe in Crisis: 1598–1648* (Malden, MA: Blackwell, 2001) 4–5.

31 Ibid., 4–8.

32 Joel Mokyr, "Review: The Great Conundrum," *The Journal of Modern History* Vol. 62, No. 1 (March 1990) 78.

33 Jack A. Goldstone, "East and West in the Seventeenth Century: Political Crises in Stuart England, Ottoman Turkey, and Ming China," *Comparative Studies in Society and History* Vol. 30, No. 1 (January 1988) 106. See also, Conrad Totman, *The Green Archipelago: Forestry in Pre-Industrial Japan* (Athens, OH: Ohio University Press, 1998) 172.

34 Mokyr, "Review: The Great Conundrum,"80.

35 Jack A. Goldstone, "The Demographic Revolution in England: A Re-examination," *Population Studies* Vol. 49 (1986) 5–33.

36 John Komlos, "Nutrition, Population Growth, and the Industrial Revolution in England," *Social Science History* Vol. 14, No. 1 (Spring 1990) 71–4.

37 Ibid., 80–1.

38 Ibid., 82–5.

39 John Bongaarts and Rodolfo A. Bulatao, "Completing the Demographic Transition," *Population and Development Review* Vol. 25, No. 3 (September, 1999) 515.

40 Thomas Robert Malthus, *Essay on the Principle of Population* (London: John Murray, 1830), in the Penguin classics edition (1982).

41 The data in the paragraphs that follow are taken from Amartya Sen, "Population: Delusion and Reality," *The New York Review of Books* Vol. XLI, No. 15 (September 22, 1994) 62–71.

42 http://www.census.gov/ipc/www/worldhis.html.

43 Navin Ramankutty, Jonathan A. Foley, and Nicholas J. Olejniczak, "Land Use Change and Global Food Production," in Ademola K. Braimoh and Paul L. G. Viek, *Land Use and Soil Resources* (Netherlands: Springer, 2008) 23–6.

44 Jared Diamond, *Collapse: How Societies Choose to Fail or Succeed* (New York: Viking Penguin, 2005) 495.

Chapter 5

1   Barney Cohen, "Urban Growth in Developing Countries: A Review of Current Trends and a Caution Regarding Existing Forecasts" (Washington, DC: National Research Council, April, 2003) 20.

2   Joel A. Tarr, *The Search for the Ultimate Sink* (Akron, OH: University of Akron Press, 1993).

3   Ibid., 8–9.

4   Joel Mokyr, *The Lever of Riches: Technological Creativity and Economic Progress* (New York: Oxford University Press, 1990) 20.

5   Ian Douglas, *The Urban Environment* (London: Edward Arnold, 1983) 2–3.

6   David Christian, *Maps of Time: An Introduction to Big History* (Berkeley, CA: University of California Press, 2004) 325.

7   Ibid., 326.

8   Ibid., 2.

9   Tertius Chandler and Gerald Fox, *3000 Years of Urban Growth* (New York: Academic Press, 1974) 362.

10 Ibid., 363.

11 Ibid.

12 V. Gordon Childe, "The Urban Revolution," *Town Planning Review* Vol. 21 (1950) 3–17.

13 Ibid.

14 Ivan Light, *Cities in World Perspective* (New York: Macmillan, 1983) 3.

15 Paul Wheatley, *The Pivot of the Four Quarters: A Preliminary Enquiry into the Origins and Character of the Ancient Chinese City* (Edinburgh, Scotland: Edinburgh University Press, 1971) 9, 225.

16 David Christian, *Maps of Time*, 269.

17 Michael Hudson, "From Sacred Enclave to Temple to City and Urban Form in Ancient Mesopotamia," in Michael Hudson and Baruch A. Levine (eds.) *Urbanization and Land Ownership in the Ancient Near East* (Cambridge, MA: Harvard University Press, 1999) 128.

18 Ibid., 129.

19 A Sumerian poet, translated S. N. Kramer, in *Ancient Near Eastern Texts Relating to the Old Testament* (ed. James B. Prichard, Princeton, NJ: Princeton

University Press, 1969, 3rd edn) 647–8. In abbreviated form, David Christian, *Maps of Time*, 295.

20   H. M. Cullen, "Climate Change and the Collapse of the Akkadian Empire: Evidence from the Deep Sea," *Geology* Vol. 28 (April 2000) 379–82.

21   Ibid., 131.

22   Ibid., 138–9, 141.

23   David Parkin and Ruth Barnes (eds.) *Ships and the Development of Maritime Technology in the Indian Ocean* (London: Routledge Curzon, 2002) 4–5.

24   Thorkild Jacobsen and Robert Adams, "Salt and Silt in Ancient Mesopotamian Agriculture," *Science* Vol. 128, No. 3334 (November 1958) 1251–7.

25   Ibid., 191.

26   Ibid., 194.

27   A. E. J. Morris, *History of Urban Form Before the Industrial Revolution* (New York: Longman Scientific & Technical, 1994) 31.

28   V. Gordon Childe, *New Light on the Most Ancient East* (United Kingdom: Taylor & Francis, revised edn 1952) 183.

29   Mortimer Wheeler, *Civilization of the Indus and Beyond* (London: Thames, 1966).

30   Paul Wheatley, *The Pivot of the Four Quarters*, 233–4.

31   Ibid., 76.

32   Ibid., 77.

33   J. M. Roberts, *The New Penguin History of the World* (London: Penguin, 2002) 484.

34   Charles L. Redman, *Human Impact on Ancient Environments* (Tucson, AZ: University of Arizona Press, 1999) 142.

35   Edith Ennen, *The Medieval Town* (New York: North-Holland, 1979) 33.

36   Paul M. Hohenberg and Lynn Hollen Lees, *The Making of Modern Europe, 1000–1950* (Cambridge, MA: Harvard University Press, 1985) 19.

37   Eric L. Jones, *The European Miracle: Environments, Economies and Geopolitics in the History of Europe and Asia* (Cambridge: Cambridge University Press, 2003) 178.

38   Hohenberg and Lees, 31.

39   Ibid.

40   Ibid., 51.

41   Ibid., 53.

42   Ibid., 77.

43   Jan de Vries, *European Urbanization 1500–1800* (Cambridge, MA: Harvard University Press, 1984) 40.

44   Ibid., 141.

45   Ibid., 70.

46   Robert B. Marks, *The Origins of the Modern World: A Global and Ecological Narrative* (Lanham, MD: Rowman & Littlefield, 2002) 137.

47   Jan de Vries, *European Urbanization 1500–1800*, 259.

Chapter 6

1   Theodore A. Wertime, "The Beginnings of Metallurgy: A New Look," *Science* Vol. 182, No. 4115 (November 1973) 880.

2   Ibid., 878.

3   Arun Kumar Biswas, *Minerals and Metals in Pre-Modern India* (New Delhi: D.K. Printworld, 2001) 61–3.

4   Katheryn M. Linduff, Han Rubin, and Sun Shuyun (eds.) *The Beginning of Metallurgy in China* (Lewiston, NY: The Edwin Mellen Press, 2000) 55.

5   This quote is taken from: http://www.geology.ucdavis.edu/~cowen/ ~GEL115/115CH4.html.

6   Ibid.

7   Paul T. Craddock, *Early Metal Mining and Production* (Washington, DC: Smithsonian Institution Press, 1995) 194.

8   Kevin Rosman, "Lead from Carthaginian and Roman Spanish Mines Isotopically Identified in Greenland Ice Dated from 600 B.C. to 300 A.D.," *Environmental Science & Technology* Vol. 31, No. 12 (1997) 3413–16.

9   "Pollution of the Caesars: Archeology – lead isotopes found in Greenland ice date to pollution from Roman lead mining pollution," *Discover* (March 1998) 45–7.

10 Cathy M. Ager and Robert G. Schmidt, *Persistence For Two Millennia Of Toxic Elements Released by Roman Metallurgical Industry, Extremadura, Spain* (Washington, DC: Smithsonian Institution, National Museum of Natural History, 2004) 67–9.

11 Tom Lugaski, Geology Project Homepage, University of Nevada, Reno, 1996 at http://www.unr.edu/sb204/geology/rome.html.

12 Jane C. Waldbaum, "The Coming of Iron in the Eastern Mediterranean: Thirty Years of Archaeological and Technological Research," in Vincent C. Pigott (ed.) *The Archaeometallurgy of the Asian Old World* (Philadelphia, PA: University of Pennsylvania Museum, 1999) 42–3.

13 Rudi Volti (ed.), "Iron," *The Facts on File Encyclopedia of Science, Technology, and Society Volume II* (New York: Facts on File, Inc., 1999) 554–6.

14 Kenneth Pomeranz, *The Great Divergence: China, Europe, and the Making of the Modern World Economy* (Princeton, NJ: Princeton University Press, 2000) 43–7.

15 Joseph Needham, *Science and Civilization in China*, Vol. 4, Part 3 (Taipei: Caves Books, 1986) 141–2.

16 Arun Kumar Biswas, "Minerals and Metals in Medieval India," in A. Rahman (ed.) *History of Indian Science, Technology and Culture, AD 1000–1800*, Vol. 111, Part 1 (New Delhi: Oxford University Press, 2000) 300–1.

17 Biswas, "Minerals and Metals in Pre-Modern India," 118–19.

18 Donald B. Wagner, *Iron and Steel in Ancient China* (New York: E.J. Brill, 1993) 405–9.

19 Donald B. Wagner, *Technology as Seen through the Case of Ferrous Metallurgy in Han China*, at http://www.staff.hum.ku.dk/dbwagner/EncIt/EncIt.html.

20 As quoted in Donald B. Wagner, *Blast furnaces in Song-Yuan China*, at http://www.staff.hum.ku.dk/dbwagner/SongBF/SongBF.pdf.

21 Frances and Joseph Gies, *Cathedral, Forge, and Waterwheel: Technology and Invention in the Middle Ages* (New York: Harper Perennial, 1995) 38–49.

22 Ibid., 62.

23 Jean Gimpel, *The Medieval Machine: The Industrial Revolution of the Middle Ages*, 2nd edn (London: Pimlico, 1992) 57.

24 Carlo M. Cipolla, *Before the Industrial Revolution: European Society and Economy, 1000–1700* (New York: W.W. Norton & Co., 1976) 162.

25 Ibid., 199–201.

26 Donald S. L. Cardwell, *Turning Points in Western Technology: A Study of Technology, Science and History* (Ann Arbor: University of Michigan Press, 1972) 14.

27 Frances and Joseph Gies, *Cathedral, Forge, and Waterwheel*, 266.

28 W. K. V. Gale, *Iron and Steel* (Harlow: Longmans, 1969) 87–8.

29 Cipolla, *Before the Industrial Revolution*, 229–30.

30 Ibid., 132–3.

Chapter 7

1 Franklin F. Mendels, "Proto-Industrialization: The First Phase of the Industrialization Process," *The Journal of Economic History* Vol. 32, No. 1 (March 1972) 241–61.

2 Walter Licht, *Industrializing America: The Nineteenth Century* (Baltimore, MD: The Johns Hopkins University Press, 1995) xvi.

3 Robert B. Marks, *The Origins of the Modern World: A Global and Ecological Narrative* (Lanham, MD: Rowman & Littlefield, 2002) 129.

4 Ibid., 127.

5 Ibid., 128–9.

6 Joel Mokyr (ed.), *The British Industrial Revolution: An Economic Perspective* (Boulder, CO: Westview Press, 1993) 100.

7 Colin McEvedy and Richard Jones, *Atlas of World Population History* (Harmondsworth: Penguin Books, 1978) 18, 171.

8 Jack A. Goldstone, "Gender, Work, and Culture: Why the Industrial Revolution Came Early to England but Late to China," *Sociological Perspectives* Vol. 39, No. 1. (Spring 1996) 5.

9   E. A. Wrigley and R. S. Schofield, *The Population History of England, 1541–1871: A Reconstruction* (Cambridge, MA: Harvard University Press 1981) 534.

10 Goldstone, "Gender, Work, and Culture," 14–15.

11 Barbara Molony, "Activism among Women in the Taisho Cotton Textile Industry," in Gail L. Bernstein (ed.) *Recreating Japanese Women, 1600–1945* (Berkeley, CA: University of California Press, 1991) 220, 225, and cited in Goldstone, "Gender, Work, and Culture," 15.

12 Konrad Specker, "Madras Handlooms in the Nineteenth Century," in Tirthankar Roy (ed.) *Cloth and Commerce: Textiles in Colonial India* (London: Sage, 1996) 215–16.

13 Sidney Pollard, "Industrialization and the European Economy," *The Economic History Review* Vol. 26, No. 4 (1973) 641.

14 Ibid., 641, 643.

15 E. J. Hobsbawm, *Industry and Empire: The Making of Modern English Society*, Vol. 2 (New York: Pantheon Books, 1967) 42.

16 Carlo M. Cipolla, *Before The Industrial Revolution: European Society and Economy, 1000–1700* (New York: W.W. Norton & Co., 1976) 96.

17 Ibid., 43.

18 Patrick Joyce, "Work," in F. M. L. Thompson (ed.) *The Cambridge Social History of Britain 1750–1950: People and Their Environment*, Vol. 2 (Cambridge: Cambridge University Press, 1990) 154–5.

19 Ibid., 132.

20 Peter N. Stearns, *European Society in Upheaval: Social History since 1800* (New York: The Macmillan Company, 1967) 112.

21 B. R. Mitchell, *British Historical Statistics* (Cambridge: Cambridge University Press, 1988) 26–7.

22 Joel A. Tarr, *The Search for the Ultimate Sink: Urban Pollution in Historical Perspective* (Akron, OH: University of Akron Press, 1997).

23 Roger Scola, *Feeding the Victorian City: The Food Supply of Manchester, 1770–1870* (Manchester: Manchester University Press, 1992) 282.

24 Charles Dickens, *Hard Times for These Times* (New York: Hurd and Houghton, 1870) 260.

25 Scola, *Feeding the Victorian City*, 282.

26 Ibid.

27 Edwin Chadwick, *Report … from the Poor Law Commissioners on an Inquiry into the Sanitary Conditions of the Labouring Population of Great Britain* (London, 1842) 269–370.

28 Hans-Joachim Voth, "Living Standards and the Urban Environment," in Roderick Floud and Paul Johnson (eds.) *The Cambridge Economic History of Modern Britain: Industrialization 1700–1860*, Vol. 2 (Cambridge: Cambridge University Press, 2004) 87.

29 Jeffrey G. Williamson, "Urban Disamenities, Dark Satanic Mills, and the British Standard of Living Debate," *The Journal of Economic History* (March, 1981) 82–3.

30 Rondo Cameron, "A New View of European Industrialization," *The Economic History Review* (February, 1985) 5.

31 Asa Briggs, *A Social History of England* (New York: The Viking Press, 1983) 186.

32 Stearns, *European Society in Upheaval*, 78.

33 Asa Briggs, *A Social History of England*, 191.

34 E. J. Hobsbawn, *Industry and Empire*, 75.

35 Joel Mokyr, "Accounting for the Industrial Revolution," in *The Cambridge Economic History of Modern Britain: Industrialization, 1700–1860*, Vol. 1 (Cambridge: Cambridge University Press, 2004) 19.

36 Ibid., 11–12.

37 Licht, *Industrializing America*, 9.

38 Ibid., 15.

39 Franklin F. Mendels, "Proto-Industrialization: The First Phase of the Industrial Process," 258–9.

40 Jonathan Prude, "Capitalism, Industrialization, and the Factory in Post-Revolutionary America," *Journal of the Early Republic* (Summer, 1996) 240.

41 Theodore Steinberg, *Nature Incorporated: Industrialization and the Waters of New England* (New York: Cambridge University Press, 1991) 23.

42 Ibid., 50.

43 Ibid.

44 Ibid., 70.

45 Marvin Fisher, "The 'Garden' and the 'Workshop': Some European Conceptions and Preconceptions of America, 1830–1860," *The New England Quarterly* Vol. 34, No. 3 (September, 1961) 313.

46 Ibid., 322.

47 Ibid.

48 Ibid., 323.

49 Steinberg, *Nature Incorporated*, 167.

50 Ibid., 205.

51 Ibid., 234.

52 Guillaume Tell Poussin, *The United States: Its Power and Progress*, translated by E. L. Du Barry (Philadelphia, 1851) 345. Cited in Marvin Fisher, "The 'Garden' and the 'Workshop': Some European Conceptions and Pre-conceptions of America, 1830–1860," 320.

53 Licht, *Industrializing America*, 104.

54 Mark Aldrich, "Determinants of Mortality among New England Cotton Mill Workers During The Progressive Era," *Journal of Economic History* Vol. XLII, No. 4 (December, 1982) 847.

55 Ibid., 849.
56 Licht, *Industrializing America*, 106.
57 Richard Sylla and Gianni Toniolo (eds.) *Patterns of European Industrialization* (London: Routledge, 1991) 84.
58 Barbara Freese, *Coal: A Human History* (New York: Penguin Books, 2003) 13.
59 Williard Glazier, "The Great Furnace of America," quoted in Joel A. Tarr (ed.) *Devastation and Renewal: An Environmental History of Pittsburgh and Its Region* (Pittsburgh, PA: University of Pittsburgh Press, 2003) 19.
60 Tarr, *Devastation and Renewal*, 19.
61 Muriel Earley Sheppard, *Cloud by Day: The Story of Coal and Coke and People* (Chapel Hill, NC: University of North Carolina Press, 1947) 2.
62 Joel A. Tarr, *The Search for the Ultimate Sink*, 390.
63 Ibid.
64 Freese, *Coal: A Human History*, 137.
65 Nicholas Casner, "Acid Mine Drainage and Pittsburgh's Water Quality," in Joel A. Tarr (ed.) *Devastation and Renewal*, 97.
66 J. R. McNeill, *Something New Under the Sun: An Environmental History of the Twentieth-Century World* (New York: W.W. Norton & Co., 2000) 315.
67 John N. Ingham, *Making Iron and Steel: Independent Mills in Pittsburgh, 1820–1920* (Columbus, OH: Ohio State University Press, 1991) 18.
68 J. R. McNeill, *Something New Under The Sun*, 310.
69 Ibid., 311.

Chapter 8

1 A paraphrased quotation from John E. Wills, Jr., "European Consumption and Asian Production in the Seventeenth and Eighteenth Centuries," in John Brewer and Roy Porter (eds.) *Consumption and the World of Goods* (New York: Routledge, 1993) 133.
2 Neil McKendrick, John Brewer and J. H. Plumb (eds.) *The Birth of Consumer Society: The Commercialization of Eighteenth-Century England* (Bloomington, IN: Indiana University Press, 1982).
3 K. N. Chaudhuri, *Trade and Civilization in the Indian Ocean: An Economic History from the Rise of Islam to 1750* (Cambridge: Cambridge University Press, 1985) 20.
4 Deborah Howard, "The Status of the Oriental Traveller in Renaissance Venice," in Gerald MacLean (ed.) *Re-Orienting the Renaissance: Cultural Exchanges with the East* (New York: Palgrave Macmillan, 2005) as reviewed by William Dalrymple, "The Venetian Treasure Hunt," *The New York Review of Books* (July 19, 2007) 29–30.

5 Jan de Vries, "Between Purchasing Power and the World of Goods: Understanding the Household Economy in Early Modern Europe," in John Brewer and Roy Porter (eds.) *Consumption and the World of Goods*, 85–132.

6 Charles L. Redman, *Human Impact on Ancient Environments* (Tucson, AZ: The University of Arizona Press, 1999) 184–5.

7 Richard C. Hoffmann, "Frontier Foods for Late Medieval Consumers: Culture, Economy, Ecology," *Environment and History* 7 (2001) 131–2.

8 Ibid., 137.

9 Ibid., 137–40.

10 McKendrick et al., *The Birth of Consumer Society*, 1.

11 John F. Richards, *The Unending Frontier: An Environmental History of the Early Modern World* (Berkeley, CA: University of California Press, 2003) 24.

12 Ibid., 96.

13 Ibid., 99–100.

14 Cissie Fairchilds, "The Production and Marketing of Populuxe Goods in Eighteenth Century Paris," in John Brewer and Roy Porter (eds.) *Consumption and the World of Goods*, 228.

15 T. H. Breen, "The Meaning of Things: Interpreting the Consumer Economy in the Eighteenth Century," in John Brewer and Roy Porter (eds.) *Consumption and the World of Goods*, 252.

16 Ibid., 253.

17 Ibid., 254.

18 Jordan Goodman, *Tobacco in History: The Cultures of Dependence* (New York: Routledge, 1993) 41.

19 David R. Montgomery, *Dirt: The Erosion of Civilizations* (Berkeley, CA: University of California Press 2007) 119.

20 Joyce Appleby, "Consumption in Early Modern Social Thought," in Lawrence B. Glickman (ed.) *Consumer Society in American History* (Ithaca, NY: Cornell University Press, 1999) 131–2.

21 Jordan Goodman, *Tobacco in History*, 140–1.

22 Ibid., 59.

23 Simon Schama, *The Embarrassment of Riches* (London: Collins, 1987) 195.

24 Jordan Goodman, *Tobacco in History*, 97.

25 Ibid., 98–9.

26 R. T. Ravenholt, "Tobacco's Global Death March," *Population and Development Review* Vol. 16, No. 2 (June 1990) 213–40.

27 Goodman, *Tobacco in History*, 105.

28 Robert N. Proctor, "Puffing On Polonium," *The New York Times* (December 1, 2006) A29. Edward P. Radford, Jr. and Vilma R. Hunt, "Polonium-210: A Volatile Radioelement in Cigarettes," *Science* Vol. 143, No. 3603 (January 1964) 247–9.

29 Ibid.

30 Goodman, *Tobacco in History*, 243.
31 Sidney W. Mintz, "Changing Roles of Food," in John Brewer and Roy Porter (eds.) *Consumption and the World of Goods*, 263.
32 Carole Shammas, "Changes in English and Anglo-American Consumption from 1550 to 1800," in John Brewer and Roy Porter (eds.) *Consumption and the World of Goods*, 183.
33 Sidney W. Mintz, *Sweetness and Power: The Place of Sugar in Modern History* (New York: Viking Penguin, 1985) 9, 13.
34 Ibid., 133–4.
35 Ibid.
36 Ibid., 73.
37 Ibid., 265.
38 Ibid., 266.
39 Robert W. Fogel, *Without Consent or Contract: The Rise and Fall of American Slavery* (New York: Oxford University Press, 1989) 21–2.
40 B. W. Higman, "The Sugar Revolution," *Economic History Review* Vol. LIII, No. 2 (2000) 213.
41 Sidney W. Mintz, "Foreword," in R. Guerra y Sanchez, *Sugar and Society in the Caribbean: An Economic History of Cuban Agriculture* (New Haven, CT: Yale University Press, 1964) xiv.
42 Richard P. Tucker, *Insatiable Appetite: The United States and the Ecological Degradation of the Tropical World* (Lanham, MD: Rowman & Littlefield, 2007) 9.
43 David Brion Davis, "Looking at Slavery from Broader Perspectives," *The American Historical Review* Vol. 105, No. 2 (April 2000) 455, 460.
44 Thomas Hobbes, *Leviathan or the Matter, Forme & Power of a Commonwealth, Ecclesiasticall and Civill, 1651*, ed. A. R. Waller (London: Cambridge University Press, 1904) 84.
45 Michael Tadman, "The Demographic Cost of Sugar: Debates on Slave Societies and Natural Increase in the Americas," *The American Historical Review* Vol. 105, No. 5 (December 2000) 1535, 1536–7.
46 Tucker, *Insatiable Appetite*, 7.
47 From Thomas Rugge's Mercurius Politicus Redivivus, November 14, 1659 in William H. Ukers, *All About Tea* (New York: Coffee and Tea Trade Journal, 1935), i, 41.
48 John E. Wills, Jr., "European Consumption and Asian Production in the Seventeenth and Eighteenth Centuries," in John Brewer and Roy Porter (eds.) *Consumption and the World of Goods*, 140–1.
49 Mark Pendergrast, *Uncommon Grounds: The History of Coffee and How It Transformed Our World* (New York: Basic Books, 1999) 17.
50 Historical Coffee Statistics, 2005, The International Coffee Organization (ICO) at http://www.ico.org/historical.asp.

51 Quoted in Mark Pendergrast, *Uncommon Grounds*, 400.

52 Pendergrast, 398.

53 A visitor to Guatemala in 1928 as quoted in Pendergrast, 399.

54 Grant McCracken, *Culture and Consumption: New Approaches to the Symbolic Character of Consumer Goods and Activities* (Bloomington, IN: Indiana University Press, 1988) 22.

55 Lizabeth Cohen, *A Consumers' Republic: The Politics of Mass Consumption in Postwar America* (New York: Alfred A. Knopf, 2003) 10.

56 Ibid., 73–4.

57 Richard M. Abrams, *America Transformed: Sixty Years of Revolutionary Change, 1941–2001* (New York: Cambridge University Press, 2006) 38–9.

58 Vaclav Smil, *Energy in World History* (Boulder, CO: Westview Press, 1994) 174.

59 Ibid., 211.

60 Ibid., 197.

61 Cohen, *A Consumers' Republic*, 123.

62 Richard M. Abrams, *America Transformed*, 38.

63 John Humphrey, Yveline Lecler, and Mario Salerno (eds.) *Global Strategies and Local Realities: The Auto Industry in Emerging Markets* (New York: St. Martin's Press, 2000) 105.

64 Ibid., 107.

Chapter 9

1 William B. Meyer, "Boston's Weather and Climate Histories," in Anthony N. Penna and Conrad Edick Wright (eds.) *Remaking Boston: An Environmental History of the City and Its Surroundings* (Pittsburgh, PA: University of Pittsburgh Press, 2009).

2 Ibid.

3 David E. Nye, "Path Insistence: Comparing European and American Attitudes Toward Energy," *Journal of International Affairs* Vol. 1 (Fall 1999) 131.

4 Vaclav Smil, *Energy in World History* (Boulder, CO: Westview Press, 1994) 132.

5 Ibid., 93.

6 Ibid., 96.

7 Ibid., 103.

8 Cited in David E. Nye, *America as Second Creation: Technology and Narratives of New Beginnings* (Cambridge, MA: The MIT Press, 2003) 118.

9 Ibid., 119.

10 Ibid.

11 Smil, *Energy in World History*, 107.

12 Nye, *America as Second Creation,* 118.
13 Peter Asmus, *Reaping the Wind* (Washington, DC: Island Press, 2001) 19.
14 Alfred W. Crosby, *Children of the Sun: A History of Humanity's Unappeasable Appetite for Energy* (New York: W.W. Norton & Co., 2006) 48.
15 David E. Nye, *Consuming Power: A Social History of American Energies* (Cambridge, MA: MIT Press, 1998) 18.
16 Asmus, *Reaping the Wind,* 25.
17 Lynn White, *Medieval Religion and Technology* (Berkeley, CA: University of California Press, 1978) 22.
18 Lewis Mumford, *Technics and Civilization* (New York: Harcourt, Brace, 1934) 117.
19 Asmus, *Reaping the Wind,* 28.
20 Sam H. Schurr and Bruce C. Netschert, *Energy and the American Economy, 1850–1975* (Baltimore, MD: Johns Hopkins University Press, 1960) 54, 485–7.
21 Ibid., 32.
22 Crosby, *Children of the Sun,* 62.
23 Ibid., 78–9.
24 Schurr, *Energy in the American Economy,* 62.
25 Ibid., 62–3.
26 Ibid., 71–2, 81.
27 Chang Jui-Te, "Technology Transfer in Modern China: The Case of Railway Enterprise (1876–1937)," *Modern Asian Studies* Vol. 27, No. 2 (1993) 281.
28 Nye, *Consuming Power,* 83–4.
29 Ibid., 88–9.
30 Richard C. Hoffmann, "Frontier Foods for Late Medieval Consumers: Culture, Economy, Ecology," *Environment and History* 7 (2001) 132. Crosby, *Children of the Sun,* 118.
31 David E. Nye, *Electrifying America: Social Meanings of a New Technology, 1880–1940* (Cambridge, MA: MIT Press, 1990) 388.
32 Ibid., 132.
33 Paul Roberts, *The End of Oil: On the Edge of a Perilous World* (Boston, MA: Houghton Mifflin, 2004) 33–4.
34 Schurr, *Energy in the American Economy,* 98–9.
35 Ibid., 106–7.
36 Ibid., 110–12.
37 Nye, *Consuming Power,* 179.
38 Ibid., 177–8.
39 Joel A. Tarr, "The Horse-Polluter of the City," in J. A. Tarr, *The Search For The Ultimate Sink: Urban Pollution in Historical Perspective* (Akron, OH: Akron University Press, 1997) 323.
40 Ibid., 324.

41 Ibid.

42 Roberts, *The End of Oil*, 155.

43 World Resources Institute, *1998–1999 World Resources: A Guide to the Environment* (New York: Oxford University Press, 1998).

44 Shurr, *Energy in the American Economy*, 115–16.

45 Christian Warren, *Brush with Death: A Social History of Lead Poisoning* (Baltimore, MD: Johns Hopkins University Press, 2001).

46 World Resources Institute: *A Guide to the Global Environment*, 63.

47 Paul Roberts, *The End of Oil*, 150.

48 Tom Koppel, *Powering the Future: The Ballard Fuel Cell and the Race to Change the World* (Toronto: John Wiley, 1999) 222.

49 Roberts, *The End of Oil*, 157.

50 Ibid., 179.

51 Crosby, *Children of the Sun*, 139.

52 Ibid., 143.

53 Richard Heinberg, *The Party's Over: Oil, War and the Fate of Industrial Societies* (Gabriola Island, BC, Canada: New Society Publishers, 2005) 149–51.

54 Ibid., 152–6.

55 Smil, *Energy in World History*, 216–17.

Chapter 10

1 The Nobel laureate and Dutch chemist Paul J. Crutzen argued in an essay titled "Geology of Mankind – The Anthropocene" that the Holocene should be renamed the Anthropocene because James Watt's steam engine (1780) changed the course of Earth history and placed humans at the center of global change on a geological scale. The article appeared in *Nature* Vol. 415 (2002) 23.

2 Richard B. Alley and Peter B. deMenocal, "Abrupt Climate Changes Revisited: How Serious and How Likely," United States Global Research Program, USGCRP Seminar, October 12, 2003.

3 Ibid.

4 IPCC, 2007: *Climate Change 2007: The Physical Science Basis. Contribution of Working Group 1 to the Fourth Assessment Report of the Intergovernmental Panel on Climate Change* [S. Solomon, D. Qin, M. Manning, Z. Chen, M. Marquis, K. B. Averyt, M. Tignor and H. L. Miller (eds.)] (Cambridge: Cambridge University Press, 2007), 463–4.

5 Ibid.

6 David Western, "Human-Modified Ecosystems and Future Evolution," *Proceedings of the National Academy of Sciences of the United States of America*, Vol. 98, No. 10 (May 8, 2001) 5458–65.

7   Charles A. Perry and Kenneth J. Hsu, "Geophysical, Archaeological, and Historical Evidence Support a Solar-Output Model for Climate Change," *Proceedings of the National Academy of Sciences of the United States of America,* Vol. 97, No. 23 (November 7, 2000) 12433–8.

8   Kenneth J. Hsu, *Climate and Peoples: A Theory of History* (Zurich, Switzerland: Orell Fussli Publishing, 2000) 88–97. Charles A. Perry and Kenneth J. Hsu, "Geophysical, Archaeological, and Historical Evidence Support a Solar-Output Model for Climate Change," *Proceedings of the National Academy of Sciences of the United States of America,* Vol. 97, No. 23 (November 7, 2000) 12433–8.

9   H. H. Lamb, *Climates of the Past, Present and Future,* Vols. I and II (London: Metheun, 1977) and by the same author, *Climate History and the Modern World* (New York: Routledge, 1982).

10   IPCC, *Climate Change 2007,* 467–8.

11   Ibid., 468.

12   C. Edward Skeen, "The Year without a Summer: A Historical View," *Journal of the Early Republic* Vol. 1, No. 1 (Spring, 1981) 51–67.

13   David Hackett Fischer, *Washington's Crossing* (New York: Oxford University Press, 2004).

14   Sebastian H. Mernild, Glen E. Liston, and Bent Hasholt, "East Greenland Freshwater Runoff to the Greenland–Iceland Norwegian Seas 1999–2004 and 2071–2100," *Hydrological Processes* (May 6, 2008); see also "Freshwater Runoff From the Greenland Ice Sheet Will More than Double by the End of the Century," *Earth and Climate* (June 12, 2008).

15   Jim Hansen, "The Threat to the Planet," *The New York Review of Books* (July 13, 2006) 13.

16   R. B. Alley, "The Younger Dryas Cold Interval as Viewed from Central Greenland," *Quaternary Science Reviews* Vol. 19 (2000) 213–26.

17   IPCC, *Climate Change 2007,* 818–19.

18   Ibid., 5.

19   Ibid., 2.

20   Radiative Forcing of Climate Change, National Academy of Science, March, 2005 and Report of the American Geophysical Union, Annual Meeting, 2003.

21   ppm (parts per million) is the ratio of the number of greenhouse gas molecules to the total number of molecules of dry air. For example, 280 ppm means 280 molecules of a greenhouse gas per million molecules of dry air. Use the same formula in calculating parts per billion (ppb) where one billion = 1,000 million.

22   IPCC, *Climate Change 2007,* 2.

23   Martin Weitzman, "Structural Uncertainties and the Value of Statistical Life in the Economics of Catastrophic Climate Change," National Bureau of Economic Research, Inc. Working paper 13490 (October, 2007).

24 Atmospheric concentrations of CH_4 have a range of 715 ppb prior to industrialization in 1750 to 1732 ppb in the early 1990s and 1774 ppb in 2005. The concentration exceeds greatly the natural range of 320 to 790 ppb in the past 650,000 years. With a confidence level of 90 percent, the IPCC Report states that the source of the increase in this greenhouse gas was anthropogenic activities, mainly fossil fuel use and agriculture. *Climate Change 2007:* 4.

25 Robert Socolow, "Stabilization Wedges: Mitigation Tools for the Next Century," Keynote Speech on Technological Options at the Scientific Symposium on Stabilization of Greenhouse Gases, "Avoiding Dangerous Climate Change" (February, 1–3, 2005), 9 pps.

Epilogue

1 Charles L. Redman, *Human Impact on Ancient Environments* (Tucson, AZ: University of Arizona Press, 1999) 90–1.

2 Robert B. Marks, *The Origins of the Modern World: A Global and Ecological Narrative* (Lanham, MD: Rowman & Littlefield, 2002) 11.

3 Ibid.

4 Gregory S. Gilbert and Stephen P. Hubbell, "Plant Disease and the Conservation of Tropical Forests," *BioScience* Vol. 46, No. 2 (1996) 104.

5 Redman, *Human Impact on Ancient Environments*, 93.

INDEX